Great Wildlife
of the Great Plains

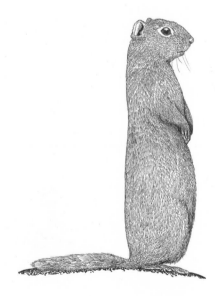

Great Wildlife

of the Great

Plains

Paul A. Johnsgard

Illustrated by the author

 University Press of Kansas

Published by the University Press of Kansas (Lawrence, Kansas 66049),
which was organized by the Kansas Board of Regents and
is operated and funded by Emporia State University, Fort Hays State
University, Kansas State University, Pittsburg State University,
the University of Kansas, and Wichita State University

Library of Congress Cataloging-in-Publication Data

Johnsgard, Paul A.
Great wildlife of the Great Plains / Paul A. Johnsgard.
p. cm.
Includes bibliographical references and index (p.   ).
ISBN 0-7006-1224-6 (cloth : alk. paper)
1. Vertebrates—Great Plains.   I. Title.
QL606.52.G74 J64 2003
596′.0978—dc21   2002012959

British Library Cataloguing in Publication Data is available.

Printed in the United States of America

10 9 8 7 6 5 4 3 2 1

The paper used in this publication meets the minimum requirements
of the American National Standard for Permanence of Paper
for Printed Library Materials Z39.48-1984.

Dedicated to all of my teachers,
whatever their guise or role.

CONTENTS

FIGURES

PREFACE

In late June 2001 I received a note from Fred Woodward, director of the University Press of Kansas, asking me if I might be interested in writing the text for a planned book on the wildlife of the Great Plains. This book was to be built around the photographs of Bob Gress, a well-known Kansas photographer and director of the Great Plains Nature Center in Wichita. As luck would have it, I was just finishing work on two other book manuscripts and had been pondering the possibility of writing a book dealing somehow with the natural history of the Great Plains region.

I immediately began organizing thoughts and references that might be relevant. I thought that a book on the "wildlife" of the Great Plains region should include at least 100 species and discuss its most characteristic breeding birds, its corresponding most typical mammals, and some of the region's more conspicuous or interesting "herps" (reptiles and amphibians)—the kinds of animals people are likely to encounter when afield with binoculars or telescopes.

Executing this conception produced a text too lengthy for the photography book and led to the Solomonic decision to offer readers two books, the first an introduction to Great Plains wildlife illustrated by my line drawings. Given the purpose of this book, then, I have excluded fishes, in part because they often aren't considered as "wildlife" but also because an excellent guide to the fishes of this same region has already been published by the University Press of Kansas (Tomelleri and Eberle 1990). The proportional essay coverage of birds (seventy-four, or 61 percent), mammals (twenty-eight, or 23 percent), and herps (nineteen, or 16 percent) may seem strong on birds, but these proportions closely reflect the fact that among the wildlife species that have been identified as Great Plains endemics there is a much larger number of species of birds (thirty-three, or 60 percent) than of mammals (sixteen, or 29 percent) or of herps (six, or 11 percent). All of these fifty-five endemics are individually discussed and nearly all are illustrated. The remaining essays describe those species that are of particular conservation importance, are particularly charismatic, or have other special attributes that make them appealing subjects.

Obviously the book could not serve as an identification guide even for the included groups; there are already a wide choice of field guides available. However, inclusion of annotated lists of the native mammals, birds, and herps that occur within the selected region seemed a useful idea.

I must thank a variety of people for their help on this project. The manuscript was read in total or in part by several friends. The mammal list was reviewed by Dr. Patricia Freeman and the herp list by Dr. Royce Ballinger, both of the Nebraska State Museum, University of Nebraska–Lincoln. I also sincerely thank two outside reviewers of the manuscript, Dr. George Potts and an anonymous reviewer. The geographic limits shown for the Great Plains region were derived from a base map kindly provided by David Wishart, editor of the forthcoming encyclopedia of the Great Plains. This map, as well as all the other maps and drawings, are by the author. The librarians of the University of Nebraska library system were as helpful as ever, and Josef Kren always promptly came to my aid when computer problems arose. The University of Nebraska has continued to support my research efforts, even after my formal retirement. I also wish to thank the University Press of Kansas for their interest in this project from the very start.

It is my hope that through a greater knowledge and appreciation of the natural heritage of the Great Plains our readership will value these fragile treasures and make personal efforts to preserve them. Many of our Great Plains species have become rare or endangered in my own lifetime, and some such as the black-footed ferret still survive by the slimmest of threads. Others like the Audubon's bighorn, gray wolf, and grizzly bear are wholly gone, at least from the Great Plains. Still others, like the plains race of the bison and the whooping crane have been rescued through concerted efforts of countless people. With perseverance, future generations might thank us for what has already been done and what we can still strive to do in the future.

"In the end, we will conserve only what we love," writes Baba Dioum. "We will love only what we understand. We will understand only what we are taught."

CHAPTER 1

The Geography and Natural Communities of the Great Plains

Few geographic regions of North America possess a greater capacity for the imagination to run free than the simple words "Great Plains": visions of bison covering the landscape from horizon to horizon, endless blue skies painted over limitless fields of grass waving hypnotically in the breeze, the smell of newly wetted black soil and fresh ozone in the air after a sky-shattering thunderstorm, and the bronzy color of bluestem and Indian grass in late fall. There is also the confident feeling that one might hike in any direction for an entire day without making detours or ever losing sight of the place you have left or the one you are headed toward. Compasses are a needless luxury in the plains. Persons who have grown up in the mountains or forests might feel slightly exposed in the plains—where can you hide from rain, hail, lightning, or tornadoes? The simple fact is, you can't. I have watched tornado funnels dance across the landscape with the ominous force of an approaching army, tossing aside everything they encounter in their paths, and have crouched helplessly on the grass while being pelted with hailstones and driving rain. But these things pass quickly; a half hour later the sun may be shining and the birds singing as if nothing at all had happened.

The Great Plains of North America both begin and end rather indefinitely. Unlike the major North American mountains or its coastlines, for which quite precise lines can be drawn on a map, the Great Plains are more notable for what they are not than what they are. At

1

their northern extremes, the Great Plains merge with the arctic coastal plain of northwestern Canada, and at their southern edge they blend with the coastal plain of the Gulf of Mexico. To the east, they are likewise gradually transformed into the Central Lowlands region of the Missouri, Mississippi, and Ohio River valleys. Only along their western borders, where they sometime quite suddenly encounter the Rocky Mountains, or at least the Rocky Mountain piedmont, can we establish fairly clear-cut points that tell us we have reached the end of our geographic rope (Figure 1).

The Great Plains thus end as they begin, not with a sudden bang but with a quiet whisper. To understand and enjoy them fully, one must be sensitive to whispers and expect to encounter neither forbidding mountains nor spectacular shorelines. Instead, their long, unencumbered vistas are an invitation to look inward toward the soul as well as outward toward the horizon, to look down on tiny prairie wildflowers and upward toward endless cerulean skies with equal appreciation, and to listen for the silent spirit within as well as the audible life without. If you come to the Great Plains to search for something new, you are very likely to find it, and in their quiet grassland refuges you might also be lucky enough to find yourself.

GEOGRAPHY AND PHYSIOGRAPHY

One of the best places to appreciate the vastness of the Great Plains is to stand atop Harney Peak, the highest point in the Black Hills of South Dakota and indeed in the entire five-state region. The Black Hills are the eroded remains of an upwarped dome in the earth's crust, forming a conspicuous "pimple" near the very center of the Great Plains region. At an elevation of more than 7,200 feet, Harney Peak commands a view of three states. Its pre-Cambrian rock core goes back to an early phase in the history of the earth, and the surrounding countryside is sequentially littered with Paleozoic, Mesozoic, and Cenozoic sediments, like an enormous geological column that has been laid out horizontally on the landscape for all to see. Beyond the ponderosa pine forests of the Hills, the short-grass prairies stretch to the very horizon. No wonder the Oglala, Brule, and other subtribes of the Teton Sioux regarded Harney Peak as sacred, for before them they could panoramically see virtually their entire high

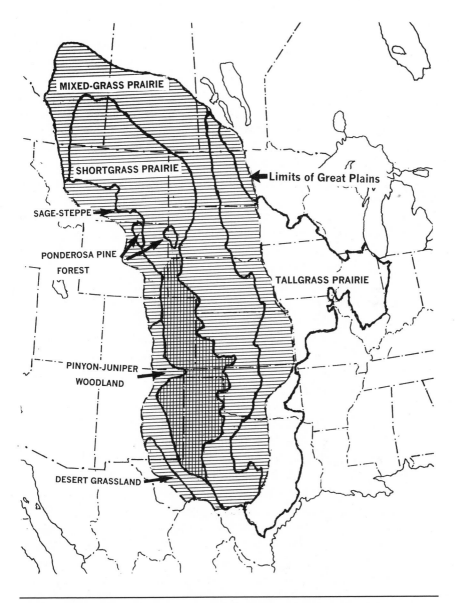

MIXED-GRASS PRAIRIE

SHORTGRASS PRAIRIE

◀—Limits of Great Plains

SAGE-STEPPE

PONDEROSA PINE
FOREST

TALLGRASS PRAIRIE

PINYON-JUNIPER
WOODLAND

DESERT GRASSLAND

Figure 1. Major native vegetation types of the Great Plains and their historic distributions (various sources). The hatched area encompasses the Great Plains region (Wishart forthcoming); cross-hatching indicates the primary short-grass region.

plains domain, the land of the bison Tatanka, the elk Hehaka, the grizzly bear Mato, the wolf Shunka, and the deer Tacha, as well as perceive their invisible worldview as encompassed in their unifying spiritual concept of Wakan-Tanka.

For the purposes of this book, the extent of the Great Plains region south of Canada is conveniently defined as being limited by the five plains states extending from North Dakota through Oklahoma, plus the Texas panhandle. This five-state-plus region encompasses some 333,000 square miles and represents about 11 percent of the area of the continental United States south of Canada. It includes parts or all of eleven physiographic regions (Figure 2), as defined by land forms, and thirteen biological regions (Figure 3), named for a combination of land types and native plant communities. It also includes most of the surviving natural prairie grasslands located south of Canada and east of the Rocky Mountains; the eastern extensions of the tallgrass prairie into Iowa, Illinois, and a few points farther east have long since been destroyed by agriculture. The arid steppes and shrub-steppes of eastern Montana, eastern Wyoming, and eastern Colorado are excluded; these areas include no grassland-dependent terrestrial vertebrates beyond those occurring within the selected region, although some sage-dependent species such as the sage sparrow and green-tailed towhee do occur. The breeding birds of this region were documented in one of my earlier books (1986).

The defined geographic limits employed in the present book are about a third less than those defined in my previous (1979) book on the breeding birds of the Great Plains, which additionally includes portions of western Minnesota and Iowa and of eastern Colorado and New Mexico. Within my currently chosen geographic limits, about 280 species of birds breed fairly regularly. Of these, over 90 percent reach their range limits somewhere within the total region. A book by J. Knox Jones Jr., David Armstrong, and Jerry Choate on the mammals of the plains states (1985) covers the same five-state region but excludes the Texas panhandle. A total of 138 native mammal species occur here, of which 90 percent likewise reach their range limits somewhere within the circumscribed region. Together, these two book references provide more details on the breeding birds and native mammals that can be offered here, and they should be consulted for details on such aspects as geographic distributions and biology. No comparable reference exists for the third group of wildlife

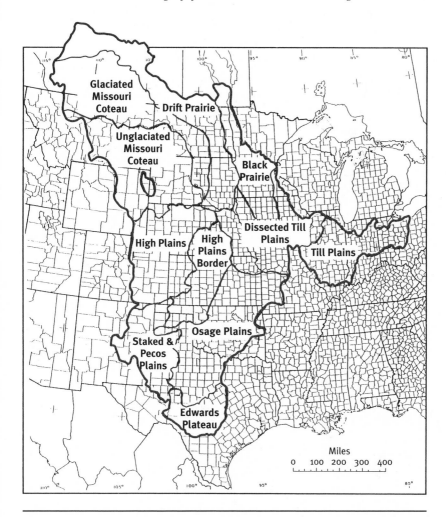

Figure 2. Physiographic regions of the Great Plains and Central Lowlands (Johnsgard 2001b)

discussed here, the reptiles and relatively terrestrial amphibians. However, several good state monographs do exist for these "herptiles" and these are cited in the references.

Although there are minor variations, the overall topography of this region is an inclined plain, which slopes downward from the west to the east at an average gradient of about ten feet per mile. Over nearly

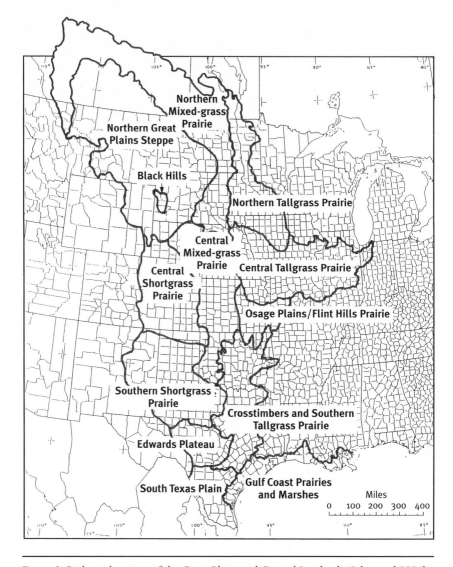

Figure 3. Biological regions of the Great Plains and Central Lowlands (Johnsgard 2001b)

6

the entire area, drainage is to the southeast, into the Missouri and Mississippi River systems. However, in North Dakota, the Souris and Red Rivers are part of the Hudson Bay drainage system, since there is an inconspicuous but important north-south continental divide in extreme northeastern South Dakota, at the southern terminus of glacial Lake Agassiz.

The highest point in the region is Harney Peak, in the Black Hills, 7,342 feet above sea level, and the lowest point is in southeastern Oklahoma's coastal plain, 323 feet above sea level. Along the eastern limits, the only highlands of significance are the Ouachita Mountains of southeastern Oklahoma, which are a western extension of the Ozark Plateau and whose sandstone ridges attain a maximum height of nearly 3,000 feet. Approximately the northern half of this region has been markedly affected by glaciers; the Wisconsinian glaciation was most recent. However, the much earlier Nebraskan and Kansan glaciations extended the farthest south, the latter reaching northeastern Kansas (Figure 4).

NATURAL VEGETATION AND FAUNAL AFFINITIES

For most people, flying over the Great Plains at 30,000 feet and peering down onto a vast flatland is to experience an understanding of the term "flyover country"; no snow-capped purple-mountain majesties or rocky coastlines delight the eye. Grasslands are never best appreciated from afar; distance might well lend enchantment to mountains, but prairies require close and personal contacts to deliver their subtle brand of magic. It is better to simply walk among them, dragging your fingers through the soft tassels of bluestem, letting its golden pollen gild your clothes, imagining that you are surrounded by endless prairie grasses and countless bison. Imagination is required—such a scene no longer really exists.

Although enormous changes have occurred that have profoundly affected the presettlement vegetation of the region, numerous historical records and sufficient relict communities still exist to provide a reasonable basis for mapping the original distribution of vegetation types through the region. Largely on the basis of a map assembled by A. W. Küchler (1966), it is possible to estimate the relative abundance of major plant communities that once covered the land surface of the region (Figure 5). Using such criteria, it seems likely that the approximate

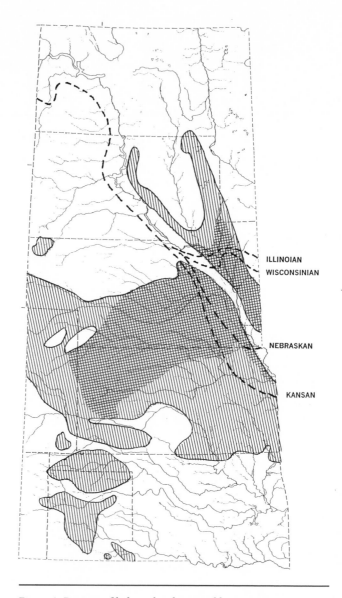

ILLINOIAN
WISCONSINIAN

NEBRASKAN

KANSAN

Figure 4. Regions of lighter (hatching) and heavier (cross-hatching) loess deposition and southern limits of major glaciations in the Great Plains states (various sources)

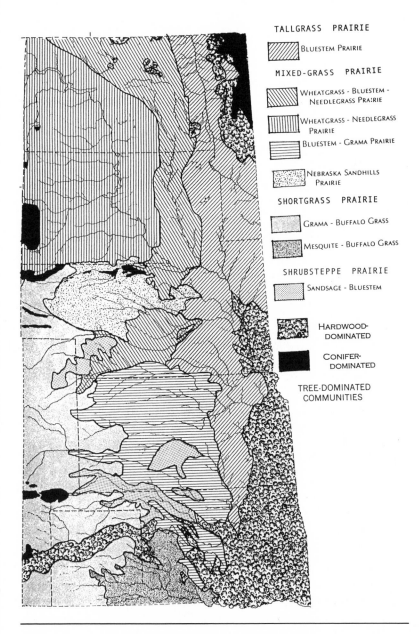

TALLGRASS PRAIRIE

Bluestem Prairie

MIXED-GRASS PRAIRIE

Wheatgrass - Bluestem -
Needlegrass Prairie

Wheatgrass - Needlegrass
Prairie

Bluestem - Grama Prairie

Nebraska Sandhills
Prairie

SHORTGRASS PRAIRIE

Grama - Buffalo Grass

Mesquite - Buffalo Grass

SHRUBSTEPPE PRAIRIE

Sandsage - Bluestem

Hardwood-
Dominated

Conifer-
Dominated

TREE-DOMINATED
COMMUNITIES

Figure 5. Historic distribution of native plant communities in the Great Plains states. Minor gallery forest extensions into grassland regions are not shown (adapted from Küchler 1966).

9

500,000-square-mile area shown in the accompanying maps was once 81 percent native grasslands, 13 percent hardwood deciduous forest or forest-grassland mosaic, 3 percent sage grasslands, and 2 percent coniferous forest or coniferous woodland. The remaining approximate 1 percent is now covered by surface water, predominantly of recent origin resulting from river impoundments.

The grassland-dominated communities in the area consist of several native prairie associations, ranging from tallgrass prairies to short-grass plains or steppe vegetation (Figure 4). There are thirty-three species of birds that have their historic breeding distributions centered on the Great Plains, based on various analyses (see Appendix 3), and sixteen species of mammals with similar plains-oriented distributions (Appendix 4). According to studies by J. Knox Jones, David Armstrong, Robert Hoffman, and Clyde Jones (1983), the most typical plains-adapted ("campestrian") mammals include the white-tailed jackrabbit, thirteen-lined Franklin's and Richardson's ground squirrels, black-tailed prairie dog, plains pocket gopher, olive-backed, hispid, and plains pocket mice, plains harvest mouse, northern grasshopper mouse, prairie vole, swift fox, black-footed ferret, spotted skunk, and pronghorn. In an earlier analysis, J. Knox Jones (1964) additionally included the bison as a grassland species but excluded the spotted skunk, which he classified as eastern. Of these species, at least half are declining, and several (the ferret, swift fox, pronghorn, and bison) were essentially extirpated but have been locally reestablished. Other seemingly characteristic plains species such as the coyote and gray wolf once had historic distributions that extended well beyond the Great Plains. Based on current distribution maps, I have also identified two species of snakes, two lizards, one turtle, and one toad as having distributions centering on the Great Plains (see Appendix 5).

The tallest and most species-rich of the American grasslands are the tallgrass bluestem prairies of the eastern Dakotas, western Minnesota and Iowa, and portions of eastern Nebraska and Kansas, terminating in northern Oklahoma. Robert Stewart (1975) listed four primary breeding birds—the upland plover, bobolink, western meadowlark, and Savannah sparrow—as associated with such prairies in North Dakota. Nearly all of this once-vast prairie is now gone; the roughly 5,000 square miles of the Kansas Flint Hills are the last, best remnants of it.

To the west of the bluestem prairies in the Dakotas lies the eastern mixed-grass prairie (identified as the wheatgrass-bluestem-needle-grass association in Figure 3). The dominant plants are shorter than those of bluestem prairie, but a large number of flowering forbs are also characteristic. Approximately 15 to 30 percent of it may still survive, based on some recent estimates. Robert Stewart listed eleven species of breeding birds primarily associated with this vegetation type in North Dakota, including the Baird's sparrow, chestnut-collared longspur, Sprague's pipit, and lark bunting. The western mixed-grass prairie (the wheatgrass-needlegrass association in Figure 4) occupied nearly all of North Dakota from the Missouri Valley westward and extended over more than half of South Dakota. About 60 percent of it may be still more or less intact. The vegetation is predominantly composed of shortgrass species and scattered midgrasses. From southern Nebraska south to northern Oklahoma, the bluestem–grama prairie lies, both in stature and in geographic location, between the tall bluestem prairies to the east and the drier grama–buffalo grass prairies to the west. Roughly 10 to 35 percent of it may still exist. Much of this midgrass community type is developed over windblown loess materials that mostly arrived and were deposited as rolling upland plains in the last million years, often during interglacial intervals. These soils are easily dug and make excellent burrows for pocket gophers, pocket mice, and kangaroo rats.

The short-grass prairie, or grama–buffalo grass association, occurs on localized slopes and dry exposures in the western Dakotas and over extensive portions of the region from western Nebraska southward to the Staked Plain of Texas. This "high plains" biota is adapted for considerable aridity, and its array of both plants and animals is somewhat restricted. Robert Stewart listed only the horned lark and McCown's longspur as primary short-grass forms in North Dakota. In southwestern Oklahoma and Texas, a variant of this vegetation type occurs with the inclusion of mesquite, the mesquite–buffalo grass association. An estimated 50 to 70 percent of this community type may still exist. In Texas, where the mesquite component is best developed, a fairly distinctive avifauna exists, including such distinctive breeding species as the Cassin's sparrow. Probably more than half of the historic short-grass prairies of the Great Plains still remain in mainly grassland vegetation, but most of these have been degraded beyond recognition by overgrazing and invasion of exotic weeds.

In several areas, extensive regions of sandy soil or sand dunes have greatly affected the vegetation. The largest of these is the Nebraska Sandhills region, where the vegetation is mainly widely spaced bunchgrasses, with the intervening areas either unvegetated or sparsely vegetated. Wet meadows and marshes at the bases of these hills allow for a birdlife essentially the same as that of the glaciated midgrass prairies to the north, with the long-billed curlew, upland sandpiper, and willet especially typical. In southwestern Kansas and Oklahoma, the large deposits of sand associated with the Cimarron and other river systems support a vegetation composed of sand-adapted grasses and sand sagebrush. James Rising (1974) listed the scaled quail, lesser prairie-chicken, and Chihuahuan raven as breeding birds especially typical of grassland and xeric scrub in southwestern Kansas. A somewhat similar sage-dominated community type occurs on clay soils in southwestern North Dakota and western South Dakota, where big sagebrush and silver sage grow in conjunction with short-grass vegetation and cactus. Robert Stewart listed only the greater sage-grouse, lark bunting, and Brewer's sparrow as primary characteristic species for this community type in North Dakota. Part of this region has been eroded by water and wind into badlands topography, which has its own array of typical breeding mammals and birds, such as rock wrens, common poorwills, and bushy-tailed woodrats.

Likewise, along the drainage of the Canadian River of Oklahoma and across the panhandle of Texas, a scrubby oak community dominated by low-stature and xeric-adapted oaks and little bluestem occurs, together with various deciduous shrubs and occasional low evergreens such as junipers. It may also serve as a distribution corridor between eastern deciduous forest species and those of the pinyons, junipers, and other conifers to the west. At least some of its breeding birds such as the juniper titmouse and western scrub-jay have their geographic affinities with the Great Basin. In their survey of mammals of the northern plains, J. Knox Jones, David Armstrong, Robert Hoffman, and Clyde Jones (1983) identified eight species with primary Great Basin distributional affinities among the mammals. They include the dwarf and Merriam's shrew, the long-legged and long-eared myotis, Nuttall's cottontail, Wyoming ground squirrel, northern pocket gopher, and sagebrush vole.

Communities dominated by deciduous or hardwood tree species

are diverse and are particularly abundant in the eastern and south-eastern parts of the region (Figure 4). The northern deciduous forest communities of western Minnesota and eastern North Dakota are a composite of types often dominated by oaks, maples, and basswood. Additionally, a substantial area of aspen grovelands are included, such as those in the Turtle Mountains of north-central North Dakota. The bird species associated with all these types are essentially those typical of the northeastern deciduous forests of North America, such as white-throated sparrow, Philadelphia vireo, mourning and chest-nut-sided warblers, ovenbird, and northern waterthrush.

Along the river systems of the Dakotas, Nebraska, and Kansas, a distinctive riverine or gallery forest, sometimes called the northern floodplain forest, provides an extremely important forest corridor linking eastern and western biotas. The significance of these river systems as gene-flow corridors has been established by a variety of studies on hybridization between western and eastern species of birds, such as the lazuli and indigo buntings, the black-headed and rose-breasted grosbeaks, the Bullock's and Baltimore orioles, and also the red- and yellow-shafted races of the northern flicker. Of all these east-west suture zones, those of the Platte and Niobrara Rivers in Nebraska are among the most important, for both of these rivers provide unbroken ecological gradients from eastern deciduous to Rocky Mountain piedmont coniferous forests. The middle Niobrara River has such a deciduous-coniferous transition zone that is less than 100 miles in length. Additionally, this river stretch supports many relict species of plants and animals from farther north that were isolated here in late glacial times and have somehow survived in the cool and shady canyon slopes.

From the Missouri Valley of the Nebraska-Iowa border southward, a forest type dominated by oaks and hickories tends to replace the northern floodplain forests along major river systems and also extends to the uplands in moister sites. Over much of eastern Kansas the oak-hickory forest occurs as a mosaic community with bluestem prairies, with dominance of one or the other dependent upon local conditions of soil, slope, and exposure. In southern Kansas and east-ern Oklahoma this mosaic pattern is replaced by the "cross timbers" community of oaks in extensive groves or growing singly, inter-spersed with medium-tall grasses such as bluestems and other prairie grass species. In the wetter portions of southeastern Oklahoma the forest becomes denser, and the oaks are supplemented with hickories and pines, resulting in an extensive oak-hickory-pine community in the southeastern United States and the Atlantic piedmont. Only here do such species as the brown-headed nuthatch and red-cockaded woodpecker breed. Among the mammals of these Oklahoma forests, those having eastern hardwood affinities include several bats and shrews, the eastern cottontail, eastern chipmunk, gray and fox squirrels, southern flying squirrel, white-footed mouse, eastern woodrat, woodland vole, and southern bog lemming.

Along the floodplains of the lower Arkansas and Red Rivers a distinctive moist to permanently wet southern floodplain forest also occurs, with oaks, tupelo, and bald cypress sharing dominance. In such forests a number of distinctly southern and southeastern species may be found, including several warblers, such as the Swainson's, prairie, pine, and yellow-throated, as well as the swamp-dependent anhinga. Local mammals include the southeastern myotis, marsh rice rat, and cotton mouse. Here too there are many exclusively southeastern amphibians and reptiles, including the American alligator and several salamanders and newts.

The coniferous-dominated communities of the region (Figure 4) are relatively few and distinctive. The Black Hills coniferous forest, together with the other ponderosa pine forests of southwestern North Dakota and western Nebraska, provides the most typically Rocky Mountain biota to be found in the entire region. Many of the associated breeding bird species are restricted to the Black Hills so far as this book's coverage is concerned. Those geographically associated with the Rocky Mountains include the Lewis's woodpecker, dusky and cordilleran flycatchers, American dipper, and MacGillivray's warbler. The brown creeper, ruby-crowned kinglet, and three-toed woodpecker might have their primary geographic affinities with either the western montane or the northern coniferous forests. Mammals geographically associated with the Rocky Mountain fauna, as identified by J. Knox Jones, David Armstrong, Robert Hoffman, and Clyde Jones (1983), include the yellow-bellied marmot, bushy-tailed woodrat, long-tailed vole, western jumping mouse, and bighorn sheep.

In northeastern New Mexico and the adjoining Black Mesa country of Oklahoma, a woodland community type dominated by low junipers and arid-adapted pines occurs on uplands and along dry river channels. Here the juniper titmouse, common bushtit, and pinyon jay all commonly occur, as does the green-tailed towhee at higher elevations. There is also a northern Mexican or "Chihuahuan" faunal element among the mammals of the western Great Plains. These species include two bats, the desert cottontail, black-tailed jackrabbit, spotted ground squirrel, silky pocket mouse, Ord's kangaroo rat, and western harvest mouse.

The northern coniferous forest of the Great Plains is a composite of three conifer-dominated vegetation types, including coniferous bogs, the Great Lakes spruce–fir forest, and the Great Lakes pine forest. With few exceptions, the breeding birds of all these community types are the same and are essentially those associated with the trans-Canadian boreal forest. This small intrusion of northern elements into the region is of considerable interest since they support many unique bird species. Species that are associated with northern Minnesota's coniferous or mixed coniferous-deciduous forests and that usually do not quite reach the region covered by this book include the osprey, several warblers, Swainson's thrush, and the purple finch. Boreal mammals that have entered the northern parts of the region include the arctic shrew, woodchuck, northern flying squirrel, meadow jumping mouse, and least weasel. Several additional mammals that were identified by J. Knox Jones, David Armstrong, and Jerry Choate (1985) as having such "boreomontane" affinities extend from Canada southward, mainly down the Rocky Mountain chain, but some also reach the northern periphery of our region, at least occasionally. These include three shrews, the snowshoe hare, least chipmunk, red squirrel, southern red-backed and meadow voles, marten, fisher, ermine, wolverine, lynx, and moose.

The Great Plains are thus a biological meeting place for northern, southern, eastern, and western elements, acquiring a kind of collective uniqueness simply by virtue of their central position, thereby becoming a sort of melting pot into which plants and animals have seeped from around all their edges. Like America itself, the plains represent a kind of composite or self-assembled land whose strength lies in their diversity and whose remnants must be treasured and protected, if only in fragmentary remnants and locations. Perhaps this book will aid people in appreciating the Great Plains even more and in recognizing the fact that they truly are great.

CHAPTER 2

Big Bluestem and Small Sparrows

The Tallgrass Prairie

The original tallgrass prairies once stretched downward across the plains of central North America like a slightly drawn bow. Its top end curved gracefully in a southeastward direction across the southern parts of Saskatchewan and Manitoba, following the valleys of the Red and James Rivers of North and South Dakota, continuing south through the eastern thirds of Nebraska, Kansas, and Oklahoma, and tapering to a tip in eastern and east-central Texas. To the east, across Iowa, northern Missouri, Illinois, and parts of a few peripheral states, fragments of tallgrass prairie also occurred, at least periodically, wherever and whenever the frequency of fires prevented the eastern deciduous forests from wresting dominance away from the small but fast-growing and deep-rooted perennial grasses.

Chief among these native grasses were the bluestems, especially big bluestem, which with similar-stature grasses such as Indian grass and switchgrass often reached six feet in height during late summer. Then their flowing heads lavishly cast golden pollen to the winds, lightly gilding the shaggy fuscous coats of bison that strayed through them, as the animals wandered southward with the approaching autumn. The bison are now long gone, as indeed are nearly all of the virgin tallgrass prairies. Of all the natural gifts of North America, few were more completely squandered and destroyed than our uniquely American bison, and also the native grasslands on which they depended. All that is left are a few preserves that are mostly too small to let a visitor imagine what it must have been like to see nothing but tallgrass prairie in every direction to the far horizon. In the Great Plains states from North Dakota to Texas alone, there historically may have been more than 200,000 square miles of tallgrass prairie; today no more than about 3 percent remains, and far less than 1 percent is protected by federal, state, or private agencies.

The intrusion of Euro-American explorers, the military, and finally settlers caused massive change in the Great Plains ecosystem, beginning with the exploitation and ultimately the destruction of the countless bison herds, followed by the elk and other large grazing mammals. With the disappearance of the bison, the Native American cultures of the western plains were also destined to die, as were the gray prairie wolf and the rest of the major players in the tallgrass ecosystem. Cattle now have replaced the bison, horses the elk, coyotes the wolf, and farmsteads the earthen lodges and teepees of the plains-adapted Native Americans.

Nearly all of this happened within only a few generations following the Civil War. My maternal grandmother well remembered seeing roaming "renegade" Dakota braves camping on their North Dakota homestead farm, and her parents were sent to nearby Fort Abercrombie for their safety during one of the native uprisings. But by then the bison and elk were already gone from eastern North Dakota, and most of the few remaining wolves had retreated to the remote badlands of the state's far western regions. The last one recorded near my hometown in southeastern North Dakota was tracked down and shot nine years before I was born, in 1922. My mother remembered hearing stories of wolves during the early 1900s, if not the animals themselves. She also was very familiar with the booming calls of greater prairie-chickens displaying during spring in meadows near the Sheyenne River.

The farm I knew so well as a child still exists and is adjacent to what is now the Sheyenne National Grassland, the biggest remaining parcel of federally owned tallgrass prairie in North Dakota. Greater prairie-chickens still maintain a slim and precarious foothold there, choosing this area for their last, probably futile, stand in North Dakota, and upland sandpipers still sing their fluty songs during territorial flights. Being there in late spring or summer is a quasi-religious experience for me; not only did my immediate ancestors tread these very lands, but also the rich smells and sights of native prairies simply intoxicate me, and indeed anybody who is willing to take the time needed to know and appreciate this magical place.

Greater prairie-chickens (Figure 6) could have been given a slightly more mellifluous name; perhaps prairie grouse or even prairie spirit might have been more appealing. They are indeed rather spiritlike, inasmuch as they seem to disappear completely each year immediately after their spring booming season, remaining almost invisible all summer, fall, and winter, then reappearing the following spring as soon as the melting snow begins to expose their traditional display sites. There

Figure 6. Greater prairie-chicken, adult male

is something reassuring about their certain annual regularity, and likewise something soothing and mystical about the males' main mating calls. Although called "booming," their notes have the soft texture and smoothness of a mourning dove's notes and the melancholic aspect of some lost soul seeking forgiveness. Like a distant foghorn, they might make one think of other times and other places, but for those who know them well they will cause one to want to be present in the very center of their activity, wrapped in the gilt-edged light of dawn and caught up in the sexual energy of the moment.

Of all the kinds of North American birds that have gone extinct, none has been more widely documented, perhaps because it was so closely associated with our earliest English settlers along the Massachusetts and Long Island shores, than the heath hen, a very close relative (subspecies) of the greater prairie-chicken of the tallgrass prairies. It provided abundant and easily obtained meat for these early immigrants, just as greater prairie-chickens did when settlers moved west into the plains. By the early 1930s the last surviving bird had died. In contrast, greater prairie-chickens of the eastern tallgrass prairies simply "followed the plow" west, thriving on the new sources of cultivated grains during fall and winter but depending on natural grasslands for nesting and brood-rearing during spring and summer. Eventually the percentage of cultivated croplands so far exceeded that of natural grasslands that the birds could no longer reproduce, and their populations gradually flickered out in one state and province after another. Within the first few decades of the twentieth century, they essentially disappeared from the southern Canadian plains, and by the end of the century they were nearly gone from almost everywhere east of the Missouri River. Now they are mainly limited to parts of South Dakota, Nebraska, and Kansas, and even in these strongholds they seem to be losing ground slowly.

The strategies of lek breeding are described in a later chapter with regard to the lesser prairie-chicken and greater sage-grouse, so no good purpose is achieved by repeating this story. Greater prairie-chickens also closely resemble sharp-tailed grouse in their ecological adaptations and lek-breeding behavior, and at least in east-central Nebraska there are areas where the two species are equally abundant, and they may even share the same display sites. Hybridization between the two species is rare, probably because the postures and calls of the males are sufficiently different that confusion on the part

Figure 7. Western (right) and eastern (left) meadowlarks, adult males

of females is unlikely. The males are quite undiscriminating, however, and male prairie-chickens have even been reported to try mating with a straw hat that was placed in the center of a display ground. They will certainly attempt to mate with a stuffed female mount of either species, indicating that all responsibilities for maintaining species integrity must reside with the more discriminating females. I once mounted a stuffed female on the wheelbase of a toy truck, attached a long string, and pulled it slowly across the lek toward my blind after all the males had gathered. However, females on wheels must have been too much for even male prairie-chickens to believe, and they promptly flew off.

Once mating has occurred, the males retire from the scene, probably not encountering the females again until fall flocks begin to form. Each female wanders off from the lek where she was fertilized and chooses a nesting place, if she has not already picked one out. In any case, the first egg of a clutch of up to about a dozen is likely to be laid in a matter of days, and more are laid on a near-daily basis until the clutch is completed. Only then does incubation begin. About twenty-four days later hatching should occur, unless the nest is plundered by skunks, raccoons, foxes, snakes, or any other of a host of possible enemies. Once hatching has been achieved, the chicks must survive another ten days or so before they can make their initial flights and thereby increase their chances of survival. Even so, they will be lucky if half the chicks survive until fall. Once their first winter has passed, the chances of survival improve somewhat, but like many gamebirds the annual survival rate of adults is little more than 50 percent. With those odds, few birds survive beyond their fourth or fifth year, and almost none make it to six or seven.

The greater prairie-chicken and eastern meadowlark (Figure 7) are essentially confined in the Great Plains to the tallgrass prairie region. However, the western meadowlark (Figure 7) and another prairie grouse, the plains race of the sharp-tailed grouse, are largely associated with short-grass and mixed-grass prairies. The amount of geographic and ecologic overlap between the two meadowlark species is surprisingly small, considering that the two species are essentially identical in size, beak, and foot structure and even in plumage. It is only by their very different advertising songs that males of the two can be easily distinguished. The vicinity of Lincoln, Nebraska, is unusual in that it is located in one of the few geographic areas where the two species are almost equally abundant. Here, one can often find places where the songs of both species can be heard almost simultaneously. Frequently, among the hilly glacial moraines where small reservoirs have been formed, eastern meadowlarks are to be found in the taller grasses and moister soils near the shoreline, while western meadowlarks sing from the nearby hilltops, where the grasses are lower and more scattered. Once, when I took the famous ecologist Paul Ehrlich out birding near Lincoln, we found this to be the case and also heard one bird singing an intermediate song and having a territory roughly halfway up the hillside. Whether such birds are actually hybrids cannot be determined in the field and would require close specimen examination. Intermediate songs are not at all rare around Lincoln, and it seems likely that they mostly result from opportunities of the young birds to hear both song types while they are in the process of learning their individual songs, thereby picking up some "foreign" elements in the process.

The cheery songs of eastern and western meadowlarks are among the most typical features of North American grassland habitats. Among the grassland birds seen during the first fifteen years of breeding bird surveys in the United States and Canada, the western meadowlark was reported most frequently, the horned lark was second, and the eastern meadowlark finished in a close third place. No other grass-land bird species was even half so frequently encountered as these three. No wonder so many states have chosen meadowlarks as their official state bird. Over the survey period 1966–2000, the western meadowlark has declined at an average overall annual rate of 0.6 percent on breeding bird surveys, as compared with 2.0 percent annually for the eastern meadowlark. The eastern declined regionally (Central States Region of the U.S. Fish and Wildlife Service) at an even higher annual rate of 2.42 percent. The horned lark was close behind, with an annual national decline rate of 1.9 percent. The horned lark has never been chosen for representing a single state, probably because it is not so colorful as are meadowlarks, and its rather inconspicuous if prolonged songs are often uttered while the bird is nearly out of sight in the sky.

There was a time not long ago when great fears as to the future of another tallgrass songbird, the dickcissel (Figure 8), were being expressed, partly because of habitat losses on its Great Plains breeding grounds and partly owing to massive pesticide spraying of the bird on its South American wintering grounds, where it is a serious pest to cultivated crops. Those threats have not diminished, and the overall national dickcissel population has declined substantially in recent years, the average annual rate being 1.4 percent for the period 1966–2000.

Beyond these two threats there is a third one that dickcissels in the Great Plains must deal with, and that is their great vulnerability to nest parasitism by brown-headed cowbirds. Although the birds hide their nests remarkably well, their favored breeding habitats of taller grasslands with scattered bushes or low trees correspond exactly with those of cowbirds. The brown-speckled whitish eggs of cowbirds are poor color matches for the unspotted and pale bluish eggs of dickcissels, and yet the dickcissels seemingly accept them without exception. The birds may be unable to pierce them with their beaks or drag them out of the nest, or perhaps their brooding instincts are too strong to let them abandon the nest and begin again. In any case, a higher incidence of dickcissel nests are parasitized than for any of the endemic grassland sparrows; the frequencies of one or more cowbird eggs being present in a Great Plains dickcissel nest generally range from a low of about 30 percent to as high as 90 percent. A single cowbird chick in the nest might not be fatal to the young dickcissels, but

Figure 8. Henslow's sparrow (left) and dickcissel (right), adult males

two such chicks are likely to spell disaster. A female cowbird probably never lays two eggs in the same nest, but the density of cowbirds in good dickcissel habitat is so high that parasitism by multiple females is a good possibility.

In spite of all these threats, male dickcissels never fail to sing lustily from the time they arrive in middle to latter May until midsummer. Then along country roads in eastern Nebraska it is easy to count five or six males perching on bushes and lower wires within a single country mile, even while driving at normal speed. Only mourning doves are more likely to show up in larger numbers during such superficial speed-limit surveys.

As a child in southeastern North Dakota during the 1930s, I never saw or heard dickcissels, but now they are quite common around my little hometown. Either my powers of observation were much poorer then than now, and I possessed neither binoculars nor good vision, or they have expanded their North Dakota range considerably in the past half century. Similarly, they are at the edge of their range in western Nebraska, and I didn't encounter them around our field station for several years. When I first heard them there, I wasn't even sure of their identity, for western Nebraska dickcissels have a different song style than eastern ones, and every year it takes some mental adjustment to recognize them.

The dickcissel is little more than a dressed-up sparrow, and the Great Plains grasslands are a kind of sparrow heaven; of the thirty-plus Great Plains endemic birds, roughly half are sparrows, which have all evolved stubby seed-eating beaks that can effectively gather and crush the tiny seeds typical of grasses and sedges. At least in ecological theory every sparrow species has a slightly different foraging and breeding ecology from all the others that breed in the same region—otherwise over time some would be outcompeted for the same resources and disappear. The theory is hard to prove, as ecological differences need only be statistical rather than absolute to be effective in reducing competition over time, but the large number of similar-sized and coexisting grassland sparrows would be a fertile testing ground.

Of the three grassland sparrows considered here (Figures 8 and 9), the one with the most narrow and shortest beak is the grasshopper; the Henslow's has a slightly longer and wider beak, and the vesper sparrow a slightly shorter but still somewhat wider beak, if one is counting millimeter differences. Presumably each of these features is adaptive in some way, relative to seed size, hardness, or other traits. It is well known that during the breeding season all these sparrows

Figure 9. Grasshopper (left) and vesper (right) sparrows, adult males

shift largely to a protein-rich insect diet, both for themselves and for feeding their youngsters. In the case of the grasshopper sparrow, the diet during this season is largely composed of small grasshoppers, and the other two species also include grasshoppers within their somewhat broader spectrum of summer foods.

Each of these species has a distinctive territorial or advertising song. That of the Henslow's is shortest and simplest, a brief insect-like vocalization that is not very loud but carries a considerable distance under calm conditions. The grasshopper sparrow also has an insectlike call, this one more resembling a buzzing locust or grasshopper and consisting of a prolonged trill. Lastly, the vesper's song is prolonged and musical, something like that of a song sparrow's. The species is called the vesper sparrow because it often sings in the evening hours around dusk, the traditional time for vespers.

All three species maintain rather small territories. Vespers favor more open, frequently weedy or cultivated habitats, with scattered suitable higher singing perches. The Henslow's is largely associated with denser grassy vegetation, especially that with a mat of dead grass litter. Small shrubs are attractive, but not abundant woody vegetation. Finally, grasshopper sparrows are attracted to fairly dense pastures and meadows, with only a few emergent weeds or shrubs. Although the nests of all three species are hard to find, those of grasshopper sparrows seem to be the hardest, as they are invariably extremely well hidden with grasses both around and above the nests. In spite of this highly secretive behavior, some nests of all three species are usually found by cowbirds and are parasitized, but not as frequently as are dickcissel nests in the same vicinity.

With these various threats, the national populations of all three species have declined alarmingly during the 1966–2000 interval of the breeding bird survey. The annual rate of national decline was highest in the Henslow's sparrow (6.3 percent), intermediate in the grasshopper sparrow (3.7 percent), and lowest in the vesper sparrow (0.8 percent). The estimated rate of national decline in the Henslow's sparrow was the third greatest of all 424 species analyzed. Regional trend rates were not statistically significant except for the grasshopper sparrow, which had an annual decline rate of 2.95 percent in the Central States region.

In the same way that sparrows are the common avian currency of grasslands, ground squirrels and a variety of seed-eating mice are the mammalian "small change." Thirteen-lined ground squirrels (Figure 10) are certainly one of the most adaptable of the Great Plains ground squirrels; they might just as easily be listed under the short-grass or mixed-grass prairies, and perhaps even more appropriately belong there than in the tallgrass category. Mowed areas such as golf courses or city parks in eastern parts of the plains states are among its favorite haunts, and such human influences may have allowed this ground squirrel to spread out and move into what were once tallgrass locations from short-grass habitat predecessors. The thirteen-lined species has an unusual pattern of six dark stripes alternating with seven lighter ones, each of the dark stripes broken up by a long row of pale rounded or squarish spots that produce a pelage pattern unique among ground squirrels. The tail is fairly short and only narrowly fringed with hairs. The species is relatively nonsocial, with few vocalizations, and probably relies on scent markings for announcing its presence.

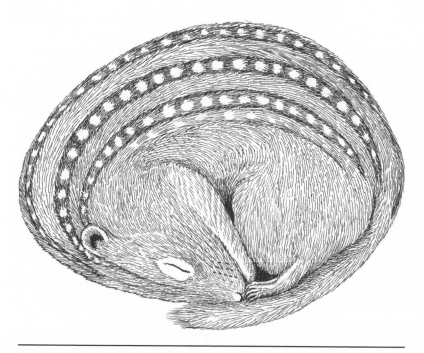

Figure 10. Thirteen-lined ground squirrel, adult hibernating

On the other hand, the Franklin's ground squirrel (Figure 11) is uniformly colored, more social, and more boreal in its geographic affinities. It was named after Sir John Franklin of Arctic exploration fame and is especially common in the transition zone between the aspen forests of southern Canada and the tallgrass and mixed-grass prairies of eastern North Dakota. Like the thirteen-lined, it is now largely associated with disturbed habitats, such as railroad rights-of-way, cemeteries, and overgrown fields. It is considerably larger and more robust than the thirteen-lined, with a longer and more luxuriant tail. It hibernates for a remarkably long period, from as early as midsummer (August) in the case of adult males. The males are soon followed into hibernation by adult females, but young of the year might remain active as late as early fall (October). Hibernation lasts until April

Figure 11. Franklin's ground squirrel, adult

or May, with adult males appearing first. Breeding begins as soon as the females emerge from hibernation, somewhat after the males. During summer the animals live in loose colonies of up to about ten to twelve animals. At that time the males maintain fairly large home ranges, which typically overlap with the smaller home ranges of several females. Territorial defense is confined to the area immediately around the burrow entrance.

Thirteen-lined and Franklin's ground squirrels are both surprisingly carnivorous, eating insects, eggs of ground-nesting birds, snakes, toads, lizards, and even young rabbits and hares. Eggs as large as duck or grouse eggs are regularly broken open and consumed, the animal curling its body around the eggs to provide a firm grip, then biting through the shell. Badgers, weasels, snakes, hawks, and other predators also take a large number of animals, and a variety of parasites and diseases are also significant mortality factors.

Like other ground squirrels, these two species use various vocal signals for much of their communication, such as uttering similar musical trills as warning signals. They also touch noses as a greeting or individual recognition signal. Like tree squirrels, ground squirrels typically flick their tails during aggressive interactions. Scent markings are probably also important social signals, and at least the thirteen-lined has cheek glands that are used for rubbing on objects in their environment. Both species are relatively nonsocial, especially the thirteen-lined. Females of both species have only a single litter per year, the number of young usually ranging from about five to nine. Within three or four weeks, the youngsters are weaned and begin to forage above ground, trying to put on enough fat to see them through their long hibernation. The young also gradually begin to disperse, with males usually moving farther from their natal burrows than females.

By late October the last of the thirteen-lined ground squirrels are likely to be hibernating. The young of the year are the final group to disappear below ground. At that time they must have a body weight of about 40 percent more than average if they are to survive the winter and emerge the following spring. While hibernating, the body temperature of the thirteen-lined may drop to as low as only five degrees (F.) above freezing, and only a few degrees above that of the burrow itself. The animal's breathing rate slows to about four per minute, and its rate of heartbeats drops from about 200 to about 5 per minute. All ground squirrels assume a ball-like posture during hibernation, with the head pulled down over the belly, the nose touching the pelvis, the feet drawn up on each side of the head, and the tail pulled over the head. There is little if any response to touching during deep hibernation. Unless the animals are in excellent physical condition at the onset of hibernation, the physical stress associated with undergoing such a long period of torpor may be fatal.

In contrast to the highly visible ground squirrels, prairie voles are among our most common native rodents, and yet most people might claim to have never seen one. Often a rustling of leaves in a dense grassy meadow or the presence of small runways resembling tiny subways at ground level or slightly below it may be the only obvious signs of these inconspicuous creatures. Such surface runways are often marked by bare and packed soil, or they may be somewhat cushioned by grass clippings produced in the process of constructing the passageways. The animals live almost entirely on green stems and leaves of grasses, sedges, and forbs, supplementing these at times with roots, seeds, bark, and tubers. Overhead cover is important to their survival, but simple invisibility from above doesn't protect them from such sharp-eared or keen-nosed enemies as northern harriers, owls, coyotes, foxes, shrews, and a host of other predators. The mortality rate of these animals is so high that even producing as many as seven young per litter and several litters per year may barely keep pace. Yet, some years are much better for breeding than others, and at such times vole "plagues" may attract predators into a local area. Other years of vole scarcity may have the opposite effect and force dispersal or starvation of predator populations.

Meadow voles seem to have simple social lives but apparently do form monogamous pair bonds. In the southern parts of their range, breeding occurs during both spring and autumn, and during summer in the north. Breeding throughout the year has also been reported in some areas or under very favorable conditions associated with unusually moist soils and luxuriant plant growth. Copulation induces ovulation, and gestation lasts about three weeks. Another three weeks or so is needed to wean the young, which average about four, after which another estrus cycle may be initiated. The young are fully grown within two months. During years of maximum vole populations, their numbers might reach as many as several hundred animals per acre. Such population peaks often seem to fluctuate regularly, their numbers usually peaking at intervals of about every three years.

Like the prairie voles, chances are good that most people who have lived in the grasslands of North America may have seen the workings of pocket gophers, but not the animals themselves. Collectively the plains and northern pocket gophers extend across the entire region covered by this book, with the northern species mostly occurring to the north and west of the plains, and differing in being smaller and in having its upper incisors smooth rather than vertically grooved. Both have external furred cheek pouches that, like those of kangaroo rats and pocket mice, are used for carrying food. The upper incisors pierce the lips, producing a distinctive "buck-toothed" appearance and allowing the animal to dig its way through earth without allowing soil into its mouth. In the course of their prodigious digging behavior, pocket gophers mix and aerate soils and generate distinctive mounds of soil that are pushed to the surface. Such mounds often extend as linear paths for many yards, and their associated tunnels may be only six to ten inches below the surface. These tunnels are not the major tunnel leading to the nest chamber. The latter is considerably deeper and thus better protected from excavation. In addition to this central nesting chamber, there are also food-storage tunnels and areas for waste deposition. As much as two and a half tons of earth may be moved in a single year by an adult pocket gopher.

The burrow system of a single pocket gopher is large and complex, perhaps covering an area of up to 5,000 square feet. The animals are active all year, with their major nesting chambers built below the frost line. During that period the animals subsist mostly on stored foods, which are primarily the roots and underground stems of herbaceous plants. Because they rarely are visible above ground, their major predators are ones that are able to enter their burrows directly, such as weasels and snakes. The animals are most likely to come to the surface at night for easy foraging, when they become vulnerable to larger owls such as great horned owls. Badgers are effective at digging out the burrows more rapidly than the gophers can excavate escape routes. The animals favor loamy or sandy soils, where digging is easy. Their tunnels leading to the surface are kept plugged when not in use, and such locations are conspicuously mounded. The animals are solitary, with reproduction occurring in late winter or early spring, and there is only a single litter per year. A typical litter consists of three or four young, which are probably tended by their mother until they approach sexual maturity at the end of their first year. Recent research indicates that the taxonomy of these gophers is quite complex, and instead of a single "plains" pocket gopher, there may instead be as many as five or six species.

Over much of the plains states, the prairie rattlesnake is the species that one is literally most likely to stumble across when hiking in grasslands, and it is not shy about striking in defense. By comparison, the massasauga (Figure 12) is nearly the smallest and probably the most reclusive of our native rattlesnakes; adults rarely if ever reach three feet in length and are unlikely to stand their ground when offered an easy means of escape. Its rattles are notably short and small, and it has a distinctive pattern of nine very large scales on the top of the head. Its black eye-stripe extends well back on the side of the neck, and the body is typically covered with an attractive pattern of dark-edged diamonds and more rounded markings. Except for the rare and tiny pygmy rattler, all of the other plains rattlesnakes are larger and have distinctive scale color patterns near their terminal rattles and on the tops of their heads (Figure 12). Like other pit vipers, rattlesnakes have small pits located behind their nostrils and below and in front of each eye; these pits serve as infrared heat detectors that allow them to track warm-blooded animals. Rattlesnakes also have elliptical pupils rather than the rounded pupils typical of all nonvenomous snakes. The massasauga's name is from the Chippewa language, meaning "great river mouth," and may relate to its usual moist habitats.

The habitats of the massasauga are varied, but moist tallgrass prairies seem favored, as are open woodlands and open sage scrub. They overlap with the prairie rattlesnake in some areas and may contact the even smaller pygmy rattler in central Oklahoma. During their long period of inactivity, from October to April in the central plains, they occupy rock crevices or rodent burrows. After emerging they are active both during day and night, and at least for small mammals their venom is as likely to be lethal as that of the larger rattlesnake species. Accounts of attacks on humans suggest that a bite even by a juvenile massasauga can be extremely painful and may require hospitalization, but is unlikely to be fatal. A survey in the 1970s suggested that there were then about 1,000 rattlesnake bites per year in the United States, with perhaps 3 percent fatal and most of these

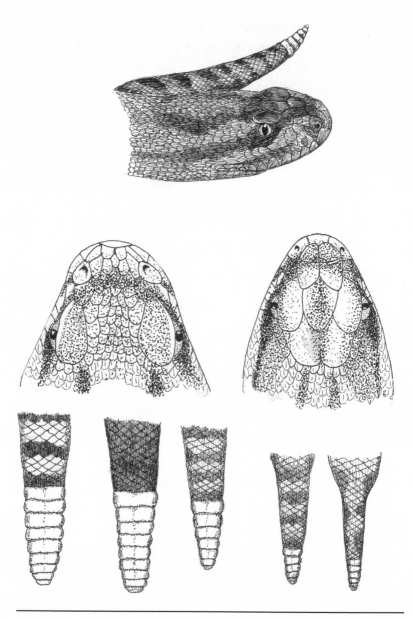

Figure 12. Massasauga (top), scale patterns of *Crotalus* (left) and *Sistrurus* snakes (middle), and (bottom, left to right) tail patterns of diamondback, timber, prairie, massasauga, and pygmy rattlesnakes

occurring in the southern states where the larger and more dangerous species such as diamondbacks are found. Few if any documented
human deaths from bites by the massasauga have been reported.
Because of their small rattles, a warning from the massasauga is less
likely to be heard by humans than those of larger species. However,
the animals are both rare and shy, and in spite of searching I have
never been able to find one in the few tallgrass prairie sites in eastern
Nebraska where they are still known to occur.

CHAPTER 3

Little Bluestem and Loess Hills

The Mixed-Grass Prairie

People lacking knowledge of physical geology often assume that the land they might happen to be standing on has always been there and furthermore will always be there. At least in the Great Plains, such is rarely the case. North and east of the Missouri River, essentially all of the present-day landscape of the Great Plains has been massively transported and reshaped by glaciers, whose telltale marks also extend southward across the eastern quarter of Nebraska and even into northeastern Kansas (see Figures 2 and 3).

Across eastern South Dakota and from roughly the southern edge of the Nebraska Sandhills south through nearly all of Kansas and locally beyond into western Oklahoma and Texas, the surface of the land has likewise been liberally sprinkled with unimaginable quantities of almost microscopic-sized particles. They had their origins in the silts deposited by meltwater streams that bordered the glaciers. These tiny materials were carried south and east for hundreds of miles from the margins of retreating glaciers during powerful windstorms of the long interglacial periods and also during the dry winters of the glacial periods themselves. The clouds of dust were eventually dropped from the sky and settled on lowlands and uplands, sometimes gradually building up into hills a hundred feet or more in depth. Because of the relative uniformity of their particle sizes, the soils that developed from these deposits had little real internal structure and instead were notably "loose." The related German term "loess" has been widely adopted by geologists to describe this distinctive type of substrate. It is not only easily tilled but also very easily eroded.

The grasslands that evolved and developed on these loess soils were among the first to be converted to small-grain agriculture. Unfortunately, they mostly occurred in regions of only moderate rainfall, where growing dryland wheat, for example, is a somewhat risky business. Their erosion-prone nature has also meant that during a century of cultivation many of these soil materials have been carried away again, either by wind or water, and have been deposited elsewhere, often via rivers into the Gulf of Mexico.

The native mixed-grass prairies of the loess hills and other substrates of the central plains mostly comprise a distinctive mixture of tallgrass species from father east and short-grass plants from the more arid, western portions of the Great Plains. As a result, these prairies average only about three feet high at maturity, and throughout their entire extent little bluestem is the single most important plant species. Unlike the sod-forming and head-high big bluestem of the tallgrass prairies, which produces a continuous grassy sward, little bluestem adopts an irregularly clumped or bunch-grass configuration. Its roots do not extend so deeply as those of big bluestem, but spread out widely into the available space around the plant, intercepting moisture near the soil surface before it can be lost to evaporation or to the thirsty roots of nearby plants. The spaces between clumps of little bluestem thus are mainly available for shallow-rooted and rapidly growing annuals or those cool-season perennials that might be able to flower and set seed quickly, before the warm-season bluestem is able to get into high gear with its powerful photosynthetic engine.

Besides the mixed-grass prairies of the loess hills, mixed grasses also came to dominate much of the glaciated areas of North Dakota and South Dakota lying to the east of the Missouri River and west of the Red and James Rivers. These prairies differ considerably from the loess-based prairies farther south, especially in their soils, which are more loamy and have greater water-holding capacity. As a result, they also have a richer forb flora. The growing season here is shorter, but the day lengths during the growing season are longer. Thus, the sequence of flowering and of breeding activity is accelerated. Because of these differences, spring arrives in the northern mixed-grass prairies with an explosion of flowering and nesting activities. More

35

species of the strictly endemic Great Plains birds, such as Baird's spar-row and Sprague's pipit, occur in the northern mixed-grass prairies of the Dakotas and adjacent eastern Montana than anywhere else on the plains.

I suppose everybody has a special plant or animal they truly asso-ciate with spring. This is especially true for those people living in the northern parts of the Great Plains, where spring's anticipated arrival is almost as painfully endured during all of the endless dreary months of winter as a woman must wait for the eventual arrival of her child. During my boyhood in North Dakota, my ultimate symbol of spring was the marbled godwit (Figure 13). Godwits didn't then (and still don't) nest in the table-flat lands of the intensively culti-

Figure 13. Marbled godwit, adult

vated clay soils of glacial Lake Agassiz where I lived, but they did seek out remnant bits of native prairie scattered along the gravelly shorelines oriented north to south just a few miles east and west of the Red River. In late April or early May, just after the last of the snow geese had passed on northward to places then unknown to me, the godwits finally arrived. They seemed to me to have all the grandeur of the infinitely more abundant snow geese, their relative rarity making their arrival all the more special. I knew that the fields they chose to nest in were ones in which I could later find such wonderful prairie plants as pasque-flowers and, still later, blazing stars.

The original English meaning of godwit was a "good thing" (to eat). Regardless of any possible eating qualities, a godwit is indeed still a very good thing (to see). They have much the same dramatic presence as a long-billed curlew, but without a decurved beak. Their beak is mostly orange during the breeding season and is sensibly almost perfectly straight, with the slightest hint of a recurved tip. Otherwise the breeding adults are dead-grass brown, save for cinnamon-tinted underwings that are evident only in flight or when the bird raises its wings vertically, as it often does during courtship display.

Wherever one finds marbled godwits, there are also likely to be upland sandpipers (Figure 14). Both are "indicator species" of native prairies, and both require large expanses of grassland to breed. In the case of upland sandpipers, areas of at least 150 acres might be needed to attract any, and even on sites as large as 500 acres their populations are likely to be only about half of their normal maximum. In very good mixed-grass habitats they may associate in loose colonies. They rarely occur in grasses more than twenty inches high or in very short grasses. They also occur in tallgrass prairies that are subjected to moderate grazing, and especially in eastern parts of their range they may nest in the mowed fields of airports.

Upland sandpipers can brighten up the landscape in a way equaled by few other birds. During the breeding season they are often heard long before they are seen. They perform high display flights, circling above their territories while occasionally uttering a loud whistle something like the "wolf whistle" sometimes used by humans to attract sexual attention. But more often the birds fly close to the ground, their wings rapidly beating in a quivering or fluttery manner, and utter a wavering whistle. It is a call certain to evoke feelings of nostalgia for the prairies in the heart of anybody who has ever experienced it.

Figure 14. Upland sandpiper, adult landing

If the upland sandpiper and marbled godwit are the hallmark birds of the mixed-grass moist meadows, then the sharp-tailed grouse is their counterpart in the uplands. The plains races of the sharp-tailed grouse (Figure 15) historically took over in the Great Plains where the greater prairie-chicken gave up for lack of winter grains. As a species, sharp-tails once ranged widely, from the open muskeg country of

Figure 15. Sharp-tailed grouse, male "dancing" display

interior Alaska and northwestern Canada southward to California, New Mexico, Kansas, and the central Great Lakes. The New Mexico and Kansas populations are now gone, as are most of the populations occurring west of the continental divide, the so-called Columbian sharp-tailed grouse. But sharp-tails still persist in the mixed-grass prairies of the Dakotas and Nebraska, only grudgingly retreating before the intensive agriculture that robs them of important winter cover and food plants. In the shrubby prairie draws they find rose hips, the buds of willows, and the seeds and fruits of a variety of plants that emerge above the snow line. During very heavy snows and times of intense cold, they often tunnel into snowbanks, where the temperature may be much warmer and where wind chills are not a factor.

Sharp-tails are, in short, tough critters and are beautiful ones as well. Their feathers are spotted above with white, as if spattered with snowflakes, and their tail feathers come to a sharp point, as if aiding in their overall streamlining. The birds are most appealing during spring courtship display, when males gather on traditional leks in the same general manner as prairie-chickens. Yet, the sounds they produce, a mixture of mechanical rattles made by tail-shaking and of squeals, coos, and yelps associated mostly with aggressive interactions, are quite different. And instead of standing about in a rather pompous manner, as do displaying greater prairie-chickens, the males motor over the ground with pattering feet, outstretched wings, and quivering tails, more resembling toy airplanes too heavy to take off than grouse engaged in serious reproductive business.

Prairie grouse time their courtship displays at dawn and dusk to avoid day-flying hawks and night-flying owls. Probably the only prairie owl likely to be active during such hours is the short-eared owl (Figure 16). On late autumn afternoons, when the sun is starting to cast long shadows across the rust-brown prairies, one sometimes suddenly sees a completely silent bird moving low over the grasses, resembling a giant moth as much as a bird, and intently going about its business of looking and listening for prairie voles or any other small rodents that make the mistake of moving even so much as a blade of grass while in its presence. Short-eared owls have very large eyes, similarly large facial disks, and enormous external ears (well hidden below the feathered facial disks) that are attuned to the slightest sounds.

In many ways the short-eared owl is an ecological counterpart species to the northern harrier, an owl-like hawk that forages in a similar manner on voles throughout the day and is replaced by the short-eared owl in the waning hours of daylight. Together they effectively

Figure 16. Short-eared owl, adult

help control small rodent populations in North American grasslands. Both are ground-nesters, an unusual trait for both owls and hawks, and adult females of both are roughly the color of dead grass. Their nests are hard to find, typically being situated in tall, thick grass, such as that growing near a wet meadow or marsh. Both of these valuable raptors have suffered greatly from habitat losses in recent decades, a pattern discouragingly shared by nearly all the grassland-dependent birds of North America. Their national population has been declining at an annual rate of 4.4 percent, the highest of any Great Plains raptor.

It has been estimated that the western meadowlark (see Figure 7) is the fourth most abundant nesting bird in North Dakota, is second (after the mourning dove) in South Dakota, and is fourth most abundant in Nebraska. Farther south, where it tends to be replaced by the eastern meadowlark, it has a much lower numerical rank as to relative abundance. Yet, it is a dearly beloved bird wherever it occurs. It is the official state bird for three of the six states (North Dakota, Nebraska, and Kansas) considered here.

In North Dakota, the western meadowlark is the earliest of the breeding songbirds to return in late winter, showing up in loose flocks that must scratch hard for food along the edges of melting snow. Soon after its arrival the males become territorial and begin their wonderful fluty songs, optimistically promising that spring is not too far off.

The ecological needs of western meadowlarks are fairly simple; grasslands as small as fifteen acres may support a meadowlark density of about half their maximum, which might only be attained in parcels of a few hundred acres or more. They prefer almost completely treeless areas, where the vegetation is about 75 percent grass and with the rest mostly consisting of forbs and a very few shrubs. The grass should be no more than about twenty inches in maximum height. Bare ground is unsuitable for nesting cover, so too-frequent burning may have undesirable effects on nesting meadowlark populations. On the other hand, occasional burning may reduce the incidence of shrubs, and only moderate grazing of taller grasslands may also have beneficial effects.

With their long and pointed beaks, meadowlarks are primarily adapted for insect-eating, and some probing for food in softer soils is possible. During fall and winter the birds shift to weed seeds and waste grain, taking whatever insects also may happen to be turned up in their searching efforts. Once they become territorial, western meadowlarks are as likely to have aggressive encounters with nearby eastern meadowlarks as with their own species. In spite of marked song differences, the males don't seem to discriminate between species. Their territories may vary from as few as two acres to as many as thirty, probably depending on habitat quality and associated population densities. Males tend toward polygyny, and some successful males may at times attract two, or rarely three, females into their own territory.

Their frequent songs, uttered on convenient posts or shrubs or sometimes during flight, are quite varied; a single male may sing as many as twelve distinct song types. Such song-switching behavior may be significant in determining a male's relative attractiveness to females as well as helping in territorial advertisement and defense. During the period 1966–2000, the western meadowlark declined an average of 0.6 percent annually as compared with 2.9 percent in the eastern meadowlark, which is more dependent on tallgrass prairies.

Probably few people who have grown up in rural areas from eastern North Dakota to central South Dakota would claim to have never seen a western meadowlark or a bobolink. Bobolink males are so colorful, and so conspicuous in spring, that they can hardly go unnoticed. Yet, most of these same people are very unlikely to admit ever seeing a Sprague's pipit, even though both species might be nesting near one another in the same meadow.

Bobolinks are named for the exuberant song they sing while on territory, a sweet "bob-o-leeee," uttered either while perched or during flight display. This flight display is performed fairly close to the ground, as the male flies about over his territory, sometimes in broad circles and occasionally landing at the point where he took off. Somewhat moist meadows seem to be preferred for breeding territories, where the grass cover is supplemented by a considerable number of broad-leaved forbs but woody cover is rare. Bobolinks prefer large grassland areas over small patches; they are rare on sites smaller than 25 acres and occur most often on prairies of about 150 acres or larger. They also like prairies with well-developed litter layers, so too-frequent hay-cropping or burning may eventually destroy their nesting habitats.

The habitat needs of Sprague's pipits are similar. They need extensive areas of standing grassy vegetation with approximate minimum area requirements of about 470 acres. They prefer native vegetation over nonnative hayfields, and many territories are located where there are high ridges with rather short grasses and a low density of forbs and sedges. But their nests are extremely well concealed in thick grasses, and few if any prairie birds have nests harder to locate than those of Sprague's pipits.

Above these wide expanses of prairie the male Sprague's pipits pour out their territorial songs while flying several hundred feet above ground. Probably no other prairie bird except perhaps horned larks flies so high during territorial flight or sings for such a prolonged period. Each bout of singing lasts up to three seconds and is followed by a series of wingbeats used to regain altitude. Then the wings are set, a long glide begins, and the song is repeated. Eventually the bird lifts or partly closes its wings and drops silently and quickly back to earth.

Nationally the Sprague's pipit declined at an annual average rate of 4.5 percent between 1966 and 2000, according to North American breeding bird surveys. The latter is one of the sharpest decline rates of all North American birds and second only to the Henslow's sparrow among prairie birds. The bobolink declined 1.6 percent annually during that same period.

Two of the many native sparrows often nesting in mixed-grass prairies are the clay-colored and Savannah sparrows (Figure 17). In truth, both have broader ecological and geographic ranges than just mixed-grass prairies. The Savannah sparrow (named for Savannah, Georgia, not a preference for savanna habitats) is a species that is broadly tolerant as to habitats, but prefers to nest in areas having intermediate vegetation density, grass and forb cover, canopy height, and litter density but with little or no woody cover present. Its breeding range extends from Alaska to Maine, in diverse herbaceous habitats. The clay-colored sparrow is more restricted in its range and ecological needs, preferring more brushy areas with scattered trees or even thickets present. Shrublands in the transition zone between prairie and aspen parklands, second-growth woods, and other distinctly woody habitats are all favored breeding sites.

Neither of these species would probably win any awards for most colorful costume. Additionally, their songs are rather forgettable; the clay-colored utters an insectlike buzzing, and the Savannah also utters a few buzzy or lisping trills. The most attractive plumage feature

Figure 17. Savannah (left) and clay-colored (right) sparrows, adult males

of the Savannah is the small yellow pattern in front of the eyes. The clay-colored has an attractive whitish stripe extending backward from each eye and a large gray nape area that extends down the sides of the neck behind the brown cheeks. Otherwise it is indeed rather generally "clay-colored."

Both of these sparrows are ground foragers, with seeds their primary food outside the nesting season. With the arrival of Savannah sparrows on nesting areas, males establish territories that may be as small as about a tenth of an acre to as large as three acres. Their boundaries tend to increase in size as the season progresses, especially if the territory-holder manages to acquire a second mate. Grasses with cover up to about forty-five inches high might be used, but lower canopies are preferred, as are sites with a good deal of litter cover and a moderate representation of forbs. Small trees may be used for singing posts when they are present. Although grasslands as small as 2.5 acres may support a territory, the birds favor larger areas, especially those larger than 25 acres.

Territories of clay-colored sparrows are surprisingly small, often much less than an acre in area, and their territories tend to be maintained from one breeding season to the next. Pair-bonds are also stronger in clay-colored sparrows than in Savannah sparrows; so far only monogamous pairings seem to have been documented. Males usually sing from perches close to the ground, and females often place their nests in the branches of a shrub. Such sites probably provide good visual protection from above as well as some shading from the afternoon sun. Brown-headed cowbirds parasitize clay-colored sparrow nests with high frequency. A lower proportion of Savannah sparrow nests have been reported parasitized, perhaps because much of the species' northern range falls outside the range of cowbirds. During the period 1966–2000, the clay-colored and Savannah sparrows declined at respective national averages of 1.1 and 0.6 percent annually on breeding bird surveys.

If the Savannah and vesper sparrows must be described as relatively drab, few if any of the grassland sparrows are more attractive than the chestnut-collared longspur (Figure 18) and Baird's sparrow. It is one of the few redeeming features of living through a North Dakota winter that nowhere else south of the Canadian prairies is there a better chance of seeing both chestnut-collared longspurs and Baird's sparrows on the same spring day, and perhaps even in the same prairie.

The two have slightly different visual appeal. The male chestnut-collared longspur is easily the Beau Brummell of prairie sparrows; its chestnut nape and white-edged ebony crown, breast, and belly markings are a stunning sight. Its song is long and musical, with up to ten

Figure 18. McCown's (left) and chestnut-collared (right) longspurs, adult males

Figure 19. White-tailed (left) and black-tailed (right) jackrabbits, adults

distinct phrases, including some repeated notes. It may last up to almost three seconds and is often performed at the apex of towering display flight that may rarely reach a height of fifty feet. Like the McCown's longspur (Figure 18) of the short-grass steppes, it then drifts slowly back to earth on outstretched wings, often while still singing loudly. Males will also sing from whatever perches that might happen to be available within their territories, which are often large glacially transported boulders, called glacial erratics.

The Baird's sparrow presents a harder basis for obtaining simple visual gratification; the primary joy of seeing a Baird's sparrow must come from the sheer pleasure of its discovery. It resembles a Savannah, clay-colored, or Henslow's sparrow because of its pale median crown stripe and a yellowish tinge above and in front of the eyes, but its breast streaking forms a distinctive dangling necklace pattern, appearing something like that of a Canada warbler. Its song is also not memorable, consisting of a few high preliminary notes, followed by a prolonged and slightly melodious trill. What the song lacks in volume and complexity, the bird makes up for in its enthusiasm, repeating the phrases every ten to fifteen seconds or so. Although seemingly simple, each male may possess several or all of twelve different song types that characterize the species. The birds are thought to be entirely monogamous, but too few pairs of Baird's sparrows have been studied to be very certain of this. It is also not very clear how frequently they are parasitized by brown-headed cowbirds; limited data suggest that the incidence may be quite high. Unfortunately, almost the entire breeding range of the Baird's sparrow falls within the zone of densest cowbird breeding populations, so probably few female Baird's sparrows escape the gaze of cowbird females while they are building their nests.

Courtship in the chestnut-collared longspur is elaborate, with wing- and tail-fanning, strutting, and exhibition of the bird's wonderful plumage patterns. In spite of its brilliant colors, the male chestnut-collared seems satisfied with a monogamous pairing system, and indeed pairs are often reestablished in the following breeding season, after a long winter hiatus. Like the Baird's sparrow, the well-hidden nests of chestnut-collared longspurs are nevertheless often found and parasitized by brown-headed cowbirds. During the period 1966–2000, the chestnut-collared longspur and Baird's sparrow declined at respective national averages of 1.6 and 2.7 percent annually on breeding bird surveys. In the Central States Region of the United States, as defined by the Fish and Wildlife Service, the Baird's sparrow declined at an average rate of 3.1 percent during the same period. The chestnut-collared longspur also seemingly declined regionally, but not quite at a statistically significant level.

Jackrabbits (Figure 19) and prairies simply belong together. In the northern parts of the Great Plains, it is the white-tailed jackrabbit that is more common. Farther south, the smaller but longer-eared black-tailed species takes over. In Nebraska both still occur fairly widely, although neither species is nearly so abundant as was once the case. Jackrabbits seem to be persecuted wherever they occur. I remember cooperative jackrabbit hunts being organized in North Dakota to try to cope with their large numbers during the dry 1930s when I was a child, but now it has been several years since I have seen either species. Their large size makes them a tempting rifle target in the

47

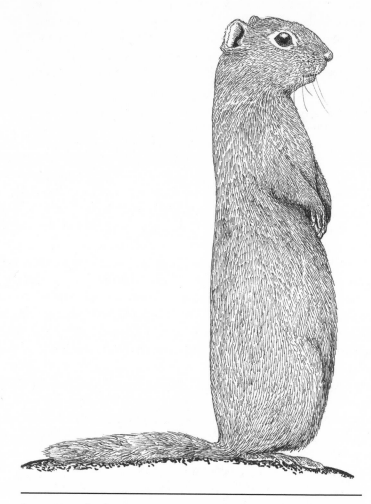

Figure 20. Richardson's ground squirrel, adult in alert "picket-pin" posture

shorter grasses where they often are still found, and they don't seem to be much more wary than cottontails, in spite of their large eyes and even larger ears. By being active mainly at night, they help to avoid notice by humans and such daytime predators as golden eagles. Coyotes are a danger to jackrabbits at any time of day or night.

Besides being larger throughout than the black-tailed, the white-tailed jackrabbit turns mostly or entirely white in the winter months, then resembling the snowshoe hare. In any season jackrabbits are strict herbivores, and probably any two species would compete strongly with one another should they occur in the same area. In such areas the two show some ecological segregation, the black-tailed often choosing the somewhat drier habitats and eating a greater diversity of foods. In the winter the animals may have to resort to eating the buds of shrubs, but otherwise grasses and forbs are their favorite foods. Hares and rabbits have remarkable feeding habits in that they regularly consume their food twice. After eating fresh vegetation, droppings of undigested food pass out the anus after a few hours and are immediately eaten again. These initial droppings are high in undigested protein and contain certain B vitamins believed to have been produced by intestinal bacteria. These materials are finally absorbed during their second pass through the animal's digestive tract.

Jackrabbits are typical hares in that, in contrast to cottontails and other true rabbits, their young are born in a relatively precocial state, with their eyes open and able to run to escape danger less than an hour after birth. The young are weaned by about a month and independent by two months. They are fully mature in less than a year but are sexually active even before then. Several litters are produced annually; in the case of the black-tailed species, it has been estimated that each female produces an average of about fourteen young per year.

Among smaller mammals, ground squirrels compose a substantial proportion of the prairie herbivore biomass. The Richardson's ground squirrel (Figure 20) locally shares its range with the thirteen-lined and Franklin's, but it is more clearly a mixed-grass prairie species than either of these. One of my memories as a very young child in North Dakota was of pouring water down a ground squirrel hole to see if we could make it emerge. I later learned that Lewis and Clark used the same technique for obtaining specimens of prairie dogs and ground squirrels. Our futile efforts occurred during the worst dust bowl days of the depression. We didn't use much water, which had to be hand-pumped and carried for some distance, so the ground squirrels might simply have appreciated the cooling they received!

These ground squirrels undergo a very long hibernation and therefore are visible above ground for only a few months, from spring melt until about July or August, depending on their age and sex. Their many predators include ferruginous, red-tailed, and Swainson's hawks, prairie falcons, badgers, weasels, coyotes, and diverse snakes,

49

to mention only the primary ones. As a result, probably few ground squirrels live longer than a year.

In North Dakota the Richardson's ground squirrel is often called a gopher, although it is not a true gopher. Many residents of the northern plains also know this species as the "flickertail," named for the quick flick of the tail used as an alarm signal, and North Dakota is still informally known as the "flickertail state." Yet, its residents have not been fond of the animals, and in earlier times the state even granted small bounties for evidence of their killing, such as their tails. During the early 1900s, the annual estimates of grain crop losses in North Dakota from this ground squirrel alone were from six to nine million dollars. Control efforts have since reduced the population of this rodent enough so that many residents of the state have probably never seen the animal by which their state is often known.

These ground squirrels are attacked by any number of predators, including hawks, owls, coyotes, weasels, and snakes. They are always on the alert for danger, often standing erect in a "picket-pin" posture, and the sight of a terrestrial predator generates a prolonged, high-pitched whistle. However, the sight of an airborne enemy causes the animal to utter a short, descending chirp and a quick retreat to its hole.

Among the many species of prairie-dwelling mice, harvest mice are distinctive in at least two ways. First, their upper incisors are distinctively grooved along their anterior surface, a curious feature for which probably every mammalogy student has given silent thanks during laboratory exams, inasmuch as it decisively separates harvest mice from all other very similar grassland mice, such as *Peromyscus*. The other feature that makes harvest mice so interesting is that they construct globular nests, somewhat like those of marsh or sedge wrens, usually quite close to the ground or even on it. The nest is typically about three inches in diameter and, like a wren's, has a small round opening located somewhere below the midpoint, often at the bottom. The inside of the nest is also somewhat birdlike but is lined with plant down rather than with feathers. In rocky ground they may also nest underneath rocks. However, upland grasslands seem to be preferred habitats.

The primary foods of harvest mice are seeds, including grains from cultivated crops, but probably more often weed seeds. A rather wide variety of larval and adult insects are also eaten. Harvest mice are strictly nocturnal and thus are rarely if ever seen by people afield during daylight hours. They are active year-round, and because they are very small they are quite sensitive to cold temperatures. As a result, they may huddle together to keep warm in their nests, even if the individuals involved are not closely related. During the night they remain quite active, perhaps in part to keep warm, and may move as far as several hundred feet during bouts of foraging. The maximum home range has been estimated at about a half acre.

In Oklahoma the breeding season is quite long, from February to November, and doubtless several litters are possible per year. One captive female produced her first litter when only three months old. The gestation period is about three weeks, and the usual litter size consists of about four young. They are weaned rapidly by about two weeks and reach adult size by five weeks. Even as adults they often weigh less than half an ounce, making them not much larger than some shrews.

Mixed-grass prairies have their share of small reptiles, including several lizards (Figure 21). Skinks are small, agile lizards that like snakes use their tongues to detect airborne odors, and probably can detect and identify members of their own and other species through olfactory signals. Yet skinks are also shiny and brilliantly colored, frequently with linear striping. The larger Great Plains skink has only a slightly striped appearance, but the prairie skink is strongly striped all the way to its tail. Skinks are active during warmer daylight hours, catching insects and scurrying to safety at the first sign of danger. They have rather short legs but long tails, and their tails are rather easily detached from their bodies when grasped, leaving the predator with only a writhing tail to eat, while the rest of the animal escapes to safety. The blood vessels in the tail stump shut down immediately and healing begins soon thereafter. The brilliant blue color of the tails of many juvenile skinks may actually serve as a distracting target for predators. In the Great Plains skink, juveniles are mostly jet black except for their blue tails, and juveniles of the prairie skink also have blue tails. This blue tail color disappears with adulthood, but males of the prairie skink develop reddish orange tints on the cheeks and chin during breeding. Great Plains skinks are surprisingly large, up to about fourteen inches long, as compared to a maximum of about eight inches in the northern prairie skink.

During the breeding season male skinks establish small territories and perform various visual displays plus apparent scent-marking with cloacal rubbing of the substrate. Fights among males may occur,

Figure 21. Collared lizard (top left), slender glass lizard (top right), Great Plains skink (middle), and prairie skink (bottom)

and courtship involves such things as head-bobbing, chin-rubbing, and other contact behavior. Skinks are egg-layers, producing a clutch of up to nearly twenty eggs that are deposited in a single place and tended by the female until they hatch about forty to fifty days later.

The slender glass lizard is often called the slender glass snake since it lacks external legs, but it also has external ear openings as well as flexible eyelids, features that are lacking in true snakes. It is called a glass lizard because, like the skinks, it has an easily detached tail; grabbing one almost anywhere will usually result in the rear part of the animal detaching and wriggling about like an earthworm while the anterior part escapes as rapidly as possible. This is a surprisingly large lizard; the adults may reach lengths of three feet, rarely even more in old males. Both sexes are strongly striped from head to tail, the tail comprising more than half of the total body length.

The slender glass lizard occupies varied habitats, but it is often found in wooded riparian situations near streams. It similarly may be common in taller grassland areas, especially near ponds. The presence of loose soil for digging may be an important habitat component. Glass lizards are mostly insectivorous but at times will eat other things, including the eggs of other lizards. In the central and southern plains, they breed in late May, the females laying a single clutch of six to seventeen eggs during June and July. They are tended by the female until hatching and then are on their own. Up to four years may be required for the young to reach sexual maturity.

CHAPTER 4

High Plains, Short Grass, and Pronghorns

The Short-Grass Prairie

In the autumn of 1999, I drove through the Pawnee National Grassland in October, on my way to Rocky Mountain National Park. I had often previously driven the main highway past the southern edge of this area on my way to the mountains and back, but never taken the time to drive into the heart of the grassland. After several years of drought, the grassland resembled a wasteland more than a short-grass prairie, rife with weedy annuals and sun-scorched grasses. Cacti were everywhere evident, and tumbleweeds were accumulating in untidy clusters along ditches and against fences. Yet, there were also birds to be seen. Several species of native sparrows were busily eating the rich crop of weed seeds, and horned larks were moving about in loose, nervous flocks, on constant alert for prairie falcons. Mountain bluebirds hopscotched from fencepost to fencepost working their way slowly southward, stopping occasionally to fly down and catch a few insects along the way. Nevertheless, the overall scene was bleak and parched. I was not sorry to leave the region and to head quickly for the snow-capped mountains that were just barely becoming visible along the western horizon.

A few months ago, this time in early June, I drove to the Rocky Mountains and, feeling hopeful, again decided to take a detour into the Pawnee National Grassland. Unlike previous years the spring had been unusually wet, and it was clear even before I entered the limits of the preserve that a remarkable transformation had occurred. In some places the landscape resembled a great flower garden rather than a grassland. White daisy fleabanes and evening primroses, purple milk-vetches, and scarlet mallows peppered the green stands of grama grasses and buffalo grass, and even the yellow flowers of prickly-pear cactus created a festive scene. Pronghorns watched warily when I slowed to a stop, and Swainson's hawks sought out the rare windmills as perching sites. In a particularly beautiful location, I stopped to photograph some flowers. While focusing on one such blossom, a movement caught the edge of my vision, and I looked up to see the black-and-white chest and throat pattern of a male McCown's longspur standing about twenty yards away on a small rock. I remained still as it eyed me with as much suspicion as I watched it with unbounded pleasure. I had searched almost in vain for longspurs on the previous trip, until finally being rewarded with a single rather unimpressive female that was feeding among a flock of horned larks along a roadside edge. In contrast, the vision of that handsome male remains indelible in my memory. One swallow may not have made a spring for perennial skeptics such as Aristotle, but a single McCown's longspur may personally commemorate a short-grass summer.

This return trip provided strong testimony to the resilience of the prairie and the capacity of native grasses to survive years of grazing, drought, even fire, and yet still return in full force when conditions improve. No doubt thousands of droughts have come and gone in the 30 million or so years since the short-grass prairies evolved on the high plains east of the Rocky Mountains, and any species unable to survive the periodic rigors of the region have long since disappeared. One of the animals that has made the North American high plains its home for the past several million years is the bison, along with its evolutionary antecedent *Bison antiquus* that gave rise to the modern species about 5,000 years ago. It eventually became the keystone species for the Native American cultures of the high plains.

For the plains-living Native Americans, the bison was at the heart of their existence. The Lakota-speaking group of the Dakota Sioux represent the Teton division of that assemblage. The Oglala and Brule subdivisions, now all living in reservations in southwestern South Dakota, had a very close, even mystical, association with bison. To them it was Tatanka, a sacred symbol as well as a functional part of their daily lives. Meat, clothing, tools, and ritual objects were all provided by the bison, and to kill a bison required both spiritual and sacrificial preparation. A piece of bison hide on a sacred pipe represented the earth itself, or the sum total of all that was visible around them. The seasonal nomadic movements of the bison generally toward the south in the fall and north in the spring meant that the people must also always be prepared to be on the move and tie their fortunes to the seasonal appearance and numbers of the buffalo. The bulls and cows each represented a variety of important virtues, such as paternal courage and maternal care, which the Oglalas themselves tried to emulate. Not only were their teepees often arranged in a circle, but the buffalo warriors also typically gathered in a circle, wearing bison headdresses and robes and mimicking the circular defensive stand of bison bulls when wolves threatened their cows and calves.

Standing somewhat below the bison in the Native Americans' worldview was the elk, Hehaka, the bulls being symbols of speed, strength, and courage. The bull elk also was believed to have mystical powers, involving long life, sexual attraction to and power over females, and similar magical traits. Like the bison, the elk was soon to disappear from the plains and retreat to the safety of the forested mountains to the west. By 1877 it was gone from Nebraska, and from South Dakota by 1888. By the early 1900s, the last survivors of the plains race of the bighorn sheep were killed in western South Dakota, the animals having disappeared some decades earlier from Nebraska and North Dakota. The skins of bighorn sheep had been especially prized for their fine wool, and when they were fashioned into dresses or shirts they were cut as little as possible to try to preserve some of the animal's spiritual powers.

Other important animals to the Oglalas and various plains tribes included mule and white-tailed deer, as the light and supple deerskin hides were perfect for clothing, bags, drumheads, and shield coverings. Wolf skins and fox skins served as ceremonial headdresses for special secret warrior societies. Likewise, otter and weasel skins provided special clothing ornaments and medicine sacks, and porcupine quills were used for decorative embroidery. Among birds, whistles made from the wing bones of eagles possessed enormous symbolic power that was transmitted to their owners in the form of great courage and endurance. Golden eagle tail and wing feathers were especially prized and considered sacred. Such feathers gave special protection to their owners, as they were somehow associated with the life-spirit *Wakan* itself.

If ever a single "keystone" animal species were to be identified with the prairie ecosystems of North America, the black-tailed prairie dog (Figure 22) would probably most easily qualify. Keystone species are those animals that tend to hold an ecosystem together, and whose presence or absence has the greatest effect on the well-being of the other species. It can be argued that predators such as the black-footed ferret, swift fox, coyote, ferruginous hawk, golden eagle, and prairie rattlesnake all largely or partly depended historically on the black-tailed prairie dog as prey. The ecological effects of prairie dogs on surrounding vegetation are exploited by the mountain plover, horned lark, lark bunting, various ground squirrels, and grasshopper mouse. Their abandoned burrows are used by burrowing owls, horned lizards, spadefoot toads, tiger salamanders, spiders, scorpions, and a variety of other invertebrates. One Oklahoma study reported fifty-six species of birds, eighteen mammals, ten reptiles, and seven amphibians using prairie dog towns. Few if any other American animals can claim so many coattail associates.

At one time in American history, the population of black-tailed prairie dogs was simply unbelievably immense, and it must have easily been the most common mammal of the short-grass plains. Some "dog-towns" in Texas were large beyond comprehension; the largest recorded one covered 25,000 square miles, and its associated animals may have numbered at least 550 million. In central Oklahoma one single town was said to extend for some twenty-two miles, and probably millions of acres were occupied in the state. Elsewhere in the plains, colonies in excess of 2,500 acres were quite common. It has been suggested that uncontrolled overgrazing by ranchers, which improved the "dog's" habitat requirements while simultaneously degrading the native grassland community, principally accounted for the spread of prairie dogs throughout the western rangelands.

Clearly this situation was bound to change, for prairie dogs and ranchers soon became bitter enemies, and the federal government in the form of the U.S. Biological Survey joined the fray on the side of the ranchers. Large-scale poisoning and fumigation programs became the watchword, and even today after more than a century of unremitting effort, the work of killing prairie dogs goes on using public funds.

Now, at most about 1,500 square miles of prairie dog habitat exist in the Great Plains, or about 1 percent of the original range of at least 150,000 square miles and a population of at least 600 million. The largest single remaining component within the plain states may be in South Dakota, where about 224 square miles of active prairie dog towns were carefully estimated in 1997–1998, as compared with 128 square miles in Nebraska and 54 square miles in North Dakota. A more recent Nebraska survey puts that state's total at closer to 50 square miles. These active remaining colonies thus constitute no more than about 0.2 percent of the total land area of the three states, hardly enough for the ranchers and state governments to declare war on the surviving animals. In 1903 it was estimated that 20 million prairie dogs occupied 2.0 to 2.5 million acres in Kansas, but by 1958 the estimate had been reduced to 57,000 acres. By 1967 there were only 280 active prairie dog colonies in Oklahoma, mainly in three counties. Most of the remaining Great Plains towns are already located within national wildlife refuges, national parks, and especially national grasslands, and only six are large enough to support a viable population of black-footed ferrets. However, political battles as to whether ranchers or prairie dogs and ferrets should have first priority in the national grasslands are far from over.

With its many diverse ecological connections, it was inevitable that the sad history of the prairie dog would be reflected in the fortunes of many other high plains species. The mountain plover, swift fox, and black-footed ferret are now all variously threatened or endangered at state or national levels. Burrowing owls have undergone precipitous declines and soon may join the list of nationally threatened species. Ironically, perhaps more money now is spent on trying to save species such as swift foxes and black-footed ferrets whose lives and fortunes were tied to prairie dogs than is wasted on trying to eliminate the last of these persistently troublesome grassland rodents.

Ranchers themselves have joined the list of threatened species in some areas; prairie dogs are only one of the many economic cards that have been dealt against them. Overgrazing of ranges, along with erosion and invasions of less edible or toxic plants, has reduced land carrying capacities, and water for livestock or forage crops is now an increasingly scarce commodity, as underground aquifers are going dry. Perhaps in the end, prairie dogs and their ecological brethren will

Figure 22. Black-tailed prairie dog, adult uttering "all-clear" call

have the final say out of economic default, at least in some especially arid regions of the Great Plains.

As a result of two legal petitions filed in 1998, the U.S. Fish and Wildlife Service concluded in 2000 that a federal listing for the prairie dog was warranted under the Endangered Species Act but was precluded owing to higher conservation priorities for other species. It is now up to the eleven affected states, including all within this book's coverage, to produce adequate state-specific conservation plans if they are to avoid federal action being taken. So far the actions of the states have been primarily to block all of the proposals that have been put forward by professional biologists.

Having lived on the plains for seventy years, it should be embarrassing to admit that I have never seen one of its most charismatic and most endemic members, the black-footed ferret (Figure 23). For much of my life I thought it already extinct, as did many biologists. Although believed to have once numbered as many as 5 to 6 million animals near the start of the twentieth century, its fate was tied to that of the prairie dog. It was last reported from Oklahoma in 1932, in Nebraska in 1949, and in North Dakota in 1980. Teddy Roosevelt helped establish the ferret's infamous reputation by describing a case in which a ferret was supposedly seen fatally clutching the throat of a pronghorn fawn and "sucking its blood with hideous greediness." He believed that the ferret preyed on all manner of birds, snakes, and mammals but correctly judged that it was the "arch enemy" of prairie dogs by following them into their burrows and thereby possibly depopulating an entire prairie dog town. It is now well known that about 90 percent of a ferret's food comes from prairie dogs alone, making its ecological tolerance extremely narrow and one that is highly dependent on a large and stable prairie dog population.

One of the great twentieth-century surprises of plains ecologists and prairie conservationists occurred when a small population of ferrets was discovered in Melette County, western South Dakota, in 1964. It was soon determined that ferrets were present in at least 20 of 151 prairie dog colonies, setting off a frantic effort to save this remnant population from extinction. Six of the animals were captured in 1971 in order to attempt captive breeding, but these soon died from unexpected responses to injections with distemper vaccine. Later capture efforts were more successful in establishing a captive stock. The South Dakota wild population disappeared by 1975, and the last sur-

Figure 23. Black-footed ferret, adult

viving captive from this group succumbed in 1979. It was again believed that ferrets had finally gone extinct.

Then, in September 1981, a plucky ranch dog near Meeteetse, Wyoming, killed a weasel-like animal that had tried to steal some of its food. Luckily, the rancher kept the carcass and took it to a local taxidermist for preservation. Even more luckily, the taxidermist recognized it as something unusual and called a conservation officer. Another chance to save the species had once again suddenly emerged. In October of that year, two biologists captured a ferret in that same general area. An intensive survey in 1982 revealed about 129 ferrets and a large prairie dog population. However, by the spring of 1985 only 52 ferrets were evident, and this population dropped alarmingly to an estimated 16 individuals by that fall as a combined result of sylvatic plague and canine distemper epidemics. Efforts to capture the survivors began in 1985, with the last known wild individual obtained in early 1987. Eighteen of these captive animals are the ancestors of all the black-footed ferrets alive today, which number in the hundreds.

Reintroductions from the progeny of captive ferrets back into the wild began in 1991. As of 2001 they have included release sites in eastern Montana (C. M. Russell National Wildlife Refuge and Fort Bellnap Indian Reservation), western South Dakota (Badlands National Park and Buffalo Gap National Grassland), eastern Wyoming (Shirley Basin), northwestern Arizona (Aubrey Valley), and the Coyote Basin of the Utah-Colorado borderland. Releases were made in the autumn of 2001 at Janus, in northern Chihuahua, Mexico, the largest known surviving prairie dog town. The Wyoming introduction failed as a result of a plague outbreak, and the same fate has befallen the Utah-Colorado release. Plague has also affected the Fort Bellnap site, but the C. M. Russell site has shown some promise.

The most successful of all releases so far has been in South Dakota's Conata Basin (Badlands National Park and adjacent Buffalo Gap grasslands), where as of 2001, over 225 ferret kits have been produced, and the free-living population was judged to be about 200 animals since initial 1994 releases. This site is the only release area fully protected from poisoning and recreational prairie dog shooting, a kill-for-the-fun-of-killing competitive activity increasingly popular in the western plains. The Bureau of Land Management has so far refused to allow ferret introductions on their vast landholdings. Instead they encourage "sportsmen" to kill as many prairie dogs as they please, using high-powered rifles and expanding bullets. This activity is euphemistically called "vaporizing" prairie dogs, whereas a much more constructive form of population control might have been provided by simply releasing ferrets. Meantime, the fate of the black-footed ferret is still in doubt.

Like the prairie dog and black-footed ferret, the story of the bison (Figure 24) is a sad one indeed. From an early estimate of some 50 to 60 million bison on the plains and prairies of central North America prior to settlement by Europeans, the situation for bison during most of the nineteenth century was one of unparalleled genocidal destruction. Until the end of the Civil War in 1865, the rate of slaughter was perhaps still controllable, but after that war there was a ready availability of improved rifles and ammunition for the general public. The completion of the transcontinental railroad in 1869 effectively split the surviving herd. It also offered easy transport for hunters heading out west and corresponding access to eastern cities for bison meat and hides being shipped back east. Although as late as 1871 a herd of

Figure 24. Bison, adult male

4 million animals still could be found in Kansas, it is likely that 4 to 6 million bison were killed from Kansas south during about three years of the early 1870s. It took only about a decade more to complete the bison's eradication, with the last remaining elements of the northern herd being tracked down in the Dakota Territory during 1873, and then completely destroyed. The last known wild bison in what is now North Dakota was killed during 1888, a year after the last one in Montana was killed and about three years after the last Nebraska survivors were eliminated from the North Platte Valley. Grizzly bears were likewise last reported in 1889 from North Dakota and about

Figure 25. Pronghorn, adult male

1890 from South Dakota. Wolverines has already disappeared from the plains by then, but gray wolves lasted somewhat longer.

Except for a small wild herd in Yellowstone Park, only a few hundred bison survived south of Canada at the start of the twentieth century, many of which were owned and managed by Native Americans. Nearly all the plains bison now alive have been derived from these and from some Canadian animals. A small group of woods-inhabiting bison also exists in northwestern Canada; this is a wild population that is often considered a distinct race and is now classified as endangered. As of 1999, nearly 300,000 bison existed worldwide, and there were roughly 200,000 bison in the United States. Most of the Great Plains bison herds are now in Montana, North Dakota, and South Dakota, and about 90 percent are privately owned. About 10,000 of the bison south of Canada are under public ownership, and nearly as many are under tribal control.

The lives of the plains bison were marked by almost constant movement; that way they never overgrazed any single area. Their periodic grazing actually stimulated the growth of new grass growth, which would become available for harvest by the next time they returned to the region. Through repetitive grazing and recurrent fires, the grass-dominated flora of the short-grass prairies slowly evolved and adapted to the high plains region, and the grazing animals in turn gradually adapted to the prairie flora and the other prairie fauna. Thus, bison are strongly attracted to prairie dog towns because of the areas of bare ground that can be used for wallows, the freshly germinating grasses and forbs caused by constant rodent grazing, and the nutritious vegetation resulting from soil and nutrient mixing associated with their digging activities.

Like most ungulates, the bison's rutting season occurs in early fall and is marked by prolonged fighting among the large and powerful bulls. Gestation lasts about nine months, so the young are born in late spring when the weather is finally improving. There is usually only a single calf, which takes about two years to attain sexual maturity. Bison are extremely fast, powerful, and should always be considered potentially dangerous to humans, especially adult males during the rut. As one who has been chased by a rutting male and somehow lived to tell about it, it is worth noting that more people have been killed recently in our national parks by bison than by grizzly bears or other more obviously dangerous mammals.

The bison may have been too large for prairie wolves to attack with impunity, but the much smaller and less dangerous pronghorn (Figure 25) must have looked like a feast ready-made for the taking. Pronghorns have often been called "antelopes," but they share only distant affinities with the true antelopes of the Old World. Instead they are exclusively a New World group, with only a single surviving species. Yet, like some small African antelopes such as gazelles, pronghorns are born to run. It is as a result of outrunning prairie predators such as wolves for uncounted millennia that every bone and muscle in a pronghorn's body has been wonderfully shaped. Along with long,

Figure 26. Swift fox, adult

slender legs, the pronghorn has also been endowed through natural selection with enormous eyes and corresponding excellent eyesight, large ears, and a fairly keen sense of smell. Their eyes are substantially larger than a human's, and the animals are reputed to detect moving humans at a distance of two or more miles, although they tend not to pay attention to completely still objects. They are able to run at a speed of up to forty-five miles per hour for prolonged periods, and for short periods may surpass fifty miles per hour or perhaps even approach sixty. In any case, they can readily outrun any other North American mammal, including horses. Even a fawn only a few days old can out-run a human and might give a coyote a very difficult chase.

If pronghorns have a serious weakness, it is in their inability or un-willingness to leap over fences or other obstacles—many pronghorns are killed in severe weather or when chased after they become en-meshed in wire fences or trapped behind similar fenced barricades. As of 2000, about 417,000 wild pronghorns existed in the Great Plains states, of which about half occurred in Wyoming. The population is now essentially stable.

Among the many unusual traits of pronghorns are their horns. Un-like the horns of cattle or the antlers of deer, the "horn" of the prong-horn consists of a horny exterior sheath grown over a bony core. Each year the exterior horny component is shed and replaced, allowing for some enlargement with age. Additionally, both sexes possess horns, although those of females are relatively small and lack prongs, or may even be entirely lacking. In immature males the small bony core of the horn is variably visible. The hairy coat of pronghorns is also unusual. The winter pelage is unusually long, with tiny air pockets in the indi-vidual hairs, providing excellent insulation. The contrasting white rear area around the rump can be erected to provide a visual alarm signal, supplemented by the release of a pungent odor. The lengthened mane can also be erected and likewise probably serves as a social alarm sig-nal. Scent glands on the head provide direct sexual stimulation to females, and territorial males often mark their territories by rubbing their faces on any available branch and bush. Their large and lustrous eyes have long eyelashes that help their vision in bright sunlight, and their eyesight notably is keen. However, they do have a sometimes fatal curiosity about unfamiliar objects seen at a distance. A fluttering handkerchief, for example, may draw an unwitting pronghorn into rifle range. And not even a pronghorn can outrun a bullet.

Pronghorns probably often provided food regularly for wolves, and their fawns are still often fair game for coyotes. But the swift fox (Figure 26) does not enter that particular branch of the prairie food chain. Compared with a coyote, the swift fox is tiny. It is at the bot-tom of a size hierarchy that goes from wolf through coyote to red fox to swift fox. An adult may weigh about 3 to 6 pounds as compared with about 6 to 8 for a red fox, 30 to 40 for a coyote, and 80 to 100 for a gray wolf. What the swift fox lacks in size, it makes up for in speed and agility. Perhaps it originally hung out around the kills of prairie wolves in the manner of present-day African jackals, hoping to find a few uneaten scraps of gristle and muscle, but always ready to dart away at a moment's notice. It probably has long been a somewhat peripheral member of the high plains predator community, and after a century or so of persecution and loss of habitat now it too has become less gristle than ghost, more memory than muscle.

Swift foxes have the same general color pattern as coyotes and other plains-dwelling canids but tend to be more tawny throughout. There is a dark patch on each side of the snout, and their tail is black-tipped, but the tail is not black along its entire upper surface as in gray foxes. Like other canids, swift foxes are monogamous, with pairs living together throughout the year but typically hunting alone, unlike wolves and coyotes. They breed at about the beginning of the year, with two to seven pups born fifty days later. They emerge from their underground den at about three or four weeks of age and soon begin spending much of their time there, playing with one another and waiting for their parents to return with food. By the time they are three months old, they begin accompanying their parents on hunts and learning the difficult art of capturing prey for themselves. Probably both adult and young are foraging opportunists; when grasshoppers locally are abundant, even they might be eaten in large quantities.

Swift foxes have suffered greatly from state and federal predator control efforts and from the loss of such basic foods as prairie dogs. Over much of their prior range they are now absent or considered as a threatened or endangered species. Reintroductions have been attempted in some of the Great Plains states and the Canadian prairie provinces. They are now apparently uncommon to rare in North Dakota, are classified as threatened in South Dakota, and as endangered in Nebraska. A few are surviving in western Kansas, where they had been judged extirpated by the early 1950s. After more than a half-century gap in records, they were rediscovered there and might now be increasing slowly. They also evidently still occur, but are very rare, in the Oklahoma and Texas panhandles. To the west, the swift fox is listed as a species of conservation concern in Montana, and it is rare in Wyoming. It is considered uncommon on the short-grass plains of northeastern Colorado, its likely primary remaining stronghold. The closely related kit fox still occurs in southwestern New Mexico, but the swift fox may now be gone from the northeastern corner of that state.

The avian predators of the short-grass prairie are numerous, and many have ecological links to prairie dogs. For example, the appearance of a ferruginous hawk (Figure 27) will convince any bird-watcher that he is indeed at home in the high plains. Although distinctly smaller than golden eagles, ferruginous hawks are actually the largest of all the North American buteos and share many of the eagle's traits. They are an open-country species that is most often seen in short-grass or sage-scrub habitats, typically wheeling high above ground, their pale pinkish tail and whitish upper-wing "window panels" often clearly visible as the bird banks in a lazy turn. They exude a sense of majesty that exactly fits their scientific name *Buteo regalis*, although a prairie dog cowering several hundred feet below in the shadow of a ferruginous hawk would probably not be thinking in such grandiose terms.

Ferruginous hawks not only have massive raptorial beaks but also unusually wide ones, allowing them to swallow good-sized rodents whole. This gives them a slightly "frog-mouthed" appearance when they open their beaks. They are also very pale on their undersides, and a ferruginous hawk directly overhead might appear almost entirely white, except for a cinnamon-toned V-mark on the lower belly that is produced by tawny leg feathering. This leg feathering actually extends down to the base of their toes, a feature that sets them and the rough-legged hawk apart from other buteos and perhaps is an adaptation for helping keep their legs warm in winter. Like red-tailed hawks the birds are prone to remain at fairly northern locations until winter, but the seasonal disappearance of prairie dogs and other hibernating rodents may eventually force them southward unless there is a good winter supply of rabbits and other prey.

The ferruginous hawk is classified as a species of special conservation concern in Kansas and Oklahoma. Yet, it is one of the very few grassland birds that has made significant population gains in recent years at the national level. In western Canada (Alberta), the population in the 1980s was no more than about half that of the 1920s, and it underwent further declines during the late 1990s.

The prairie falcon (Figure 28) is also a recipe for sudden death to prairie dogs. It is smaller than a ferruginous hawk but probably twice as fast in flight, easily reaching sixty miles an hour during an attack. Essentially the same size as a peregrine, its plumage pattern is a slightly faded and brown-toned version of the peregrine's, almost like a peregrine in grassland camouflage. It takes my breath away to see a

Figure 27. Ferruginous hawk, adult

Figure 28. Prairie falcon, adult

prairie falcon suddenly emerge from nowhere, throw a prairie dog or ground squirrel colony into utter panic, and just as quickly disappear in the distance as it appeared, apparently having decided that a better chance for a meal lies somewhere over the horizon. In such situations, lifetime memories for human observers may be forged in a matter of a few seconds, and for unlucky rodents lifetimes may be ended just as quickly.

During one memorable summer I followed the progress of a pair of prairie falcons in Grand Teton National Park. The birds were nesting on a vertical cliff overlooking a sage-covered valley supporting a large population of Wyoming ground squirrels. From their nesting vantage point, the birds could probably see every ground squirrel hole and easily detect any movement that signified a potential meal. The smaller male usually stood watch on a wind-shaped lodgepole pine at the cliff's rim, from which he could launch himself at a moment's notice to begin a low-level attack on some unwary young ground squirrel or defend the nest from any unwelcome intruders. Woe betide any raven that chanced to fly too close to the nest, for it would be lucky to escape with its life.

Prairie falcons have so far escaped the disastrous population declines of peregrines, mainly because their primary prey species have not been laced with the levels of pesticides that have so seriously affected peregrines. Yet, the birds do require large areas of wilderness or near-wilderness for their survival and without due care may well follow the peregrine into a long slide toward possible oblivion.

Mountain plovers might well have been named "prairie plovers," for except during migrations to and from their California wintering areas the birds are rarely within sight of high mountains. Like prairie falcons and ferruginous hawks, they have a somewhat bleached look about them, making them remarkably hard if not impossible to detect when they are crouching in short-grass prairie. For mountain plovers, the shorter the better seems to be the watchword in terms of grass preferences; the close-cropped soils of prairie dog colonies were probably once their prime habitat, and now they have to rely on the over-grazed pastures of marginally productive sheep or cattle ranches to find nesting sites. They are now most likely to occur where there is a substantial area of bare, relatively level ground, where the vegetation has been recently burned, where kangaroo rats or prairie dogs still survive, and where horned larks are also to be found. Not many such areas still exist in the Great Plains states, and the birds are most likely to be seen along the very western edges of the short-grass prairie.

On finally seeing a mountain plover the joy received must be largely one of simple discovery rather than the rewards of viewing a spectacularly colored prairie species such as a McCown's longspur or a lark bunting. I always think of a well-known and much-loved Quaker song, "Simple Gifts," whenever I think of the plain-plumaged mountain plover. Nationally the mountain plover population declined an average of 1.1 percent annually between 1966 and 2000, according to

North American breeding bird surveys, and it is a candidate for federal listing as a threatened species.

Not even an ecological Scrooge could easily resist the sudden appearance of a burrowing owl (Figure 29). Standing solemnly beside a prairie dog hole or perched on a nearby fencepost, the presence of a miniature owl standing erect and in plain view during the middle of the day somehow seems to be such an unlikely event that it requires a sudden stop and a prolonged look. Burrowing owls always demand my undivided attention, no matter how many times I have encountered them. There are a few other small owls in North America that are active during the day, including elf owls and pygmy owls, but the burrowing owl has a degree of panache that is hard to match. A common vernacular name, "howdy owl," seems especially appropriate, for the birds sometimes seem to provide a one-owl welcoming committee when visiting a prairie dog colony.

Although it was once believed that burrowing owls were not only welcomed by prairie dogs but might even share their burrows with them, this assumption is certainly not the case. The owls use only vacant burrows, often those at the edges of prairie dog colonies, and provide no obvious benefit to the prairie dogs. Although the owls do feed on some rodents, including some as large as thirteen-lined ground squirrels, they seem to pose no real danger to even young prairie dogs. Instead, the summer foods of burrowing owls consist largely of insects, especially scarab beetles in Nebraska, which seem to be especially common around prairie dog colonies. One can often find pellets of burrowing owls around the openings of their nesting burrows that consist exclusively of the exoskeletons of these insects. Also around the edges of the openings one is likely to find the torn-up fragments of dried cow-pies, which are of less certain significance. It is generally believed that these dried droppings may provide a sort of olfactory camouflage for the owls, but that remains to be proven, and in my experience a coyote or badger has little difficulty in selecting which burrows are active and which are abandoned. Badgers are much more of a threat to burrowing owls than are coyotes, for the digging speed and skill of badgers are phenomenal, and any owl caught in a burrow being dug out by a badger and having no "back door" is out of luck. But such an escape route is usually present, and unless two badgers are working in concert there is likely to be a safe way out, at least for fledged youngsters.

Figure 29. Burrowing owl, adult

Burrowing owls have undergone the same frightening decline in numbers as has occurred with other members of the prairie dog community. Nationally they declined an average of 1.2 percent annually between 1966 and 2000, according to North American breeding bird surveys. In western Nebraska their numbers have plummeted in close parallel with those of prairie dogs, and each year they become more difficult to locate. Martha Desmond and others observed a 63 percent reduction in burrowing owl pairs among seventeen prairie dog colonies in western Nebraska between 1990 and 1996. As prairie dog numbers decreased, there was a greater risk of predation by badgers and fewer available nesting burrows. The burrowing owl is now too rare to show up on breeding bird survey population trend analyses. The friendly "howdy owl" of the short-grass plains is now in increasing danger of having to say a final "goodbye."

The two most wonderful songbirds of the short-grass plains are the lark bunting and the McCown's longspur (see Figure 18), at least during the spring and early summer months. During fall and winter both species hardly merit more than a passing glance from anyone but the most hard-core birders, but during a high plains spring the spirited song flights by territorial males of both species can bring joy to even the most jaded spirits. Imagine driving along a dusty trail in open range country, with no trees in sight, and looking for anything that moves. Suddenly a small, mostly black bird about the size of a large sparrow erupts from the grass like a toy rocket, ascends nearly vertically upward while singing with all the volume it can muster. It stalls out about twenty feet in the air and rapidly descends, leaving the observer to wonder if the entire event had simply been an apparition. By the second or third time, it becomes clear that this is a sparrow with real pizzazz, a lark bunting.

Perhaps no sooner has one adjusted to this strange experience than another similar-sized but quite different bird also does the same thing. Its undersides are mostly white rather than black, and it has a white edging on its black tail and white underwings rather than white wing-patches on black wings. It ascends higher in the sky, up to about seventy feet, and its prolonged song is both musical and warbling rather than the somewhat staccato and less melodic output of the lark bunting. Both species gradually float back down to earth after they finish their songs, then resembling giant moths from some children's picture book.

Both species build their nests in low vegetation, often beside a small shrub or grass clump that might offer some protection from the summer sun. Females incubate for twelve days, and both sexes help feed the young, although at least sometimes male lark buntings may have two mates and may have to help raise both broods. The longspur is often double-brooded, and the lark bunting is perhaps also double-brooded, at least under favorable breeding conditions. Yet, both species have declined greatly during the thirty-five years (1966–2000) that the breeding bird survey has been monitoring their populations, the annual overall rate of decline being 1.4 percent for the lark bunting and 1.97 percent for the McCown's longspur.

The lark sparrow (Figure 30) is one of those sparrows that is just simply perfect. It has a bright, distinctive facial pattern of chestnut, white, and black that is never forgotten once it is seen, a black "stick-pin" parking on an otherwise buffy breast, and white corners on the tips of its tail that flash open in flight, allowing for easy field identification when the bird takes off. Add to these features a somewhat larklike song, a long, musical string of clear notes interrupted near the end by a few buzzy and discordant ones, like a trumpeter who nearly got to the end of his solo before making a mistake. This feature only adds to the bird's charm, making it one of the most memorable of all grassland sparrows.

The lark sparrow is also one of our more plentiful grassland sparrows, especially where there are scattered shrubs or trees to provide singing perches as well as some bare ground for easy foraging. They are increasingly common southward in the plains states, with the high plains of western Oklahoma and the Texas panhandle their center of abundance. There a profusion of weed seeds and insects such as grasshoppers and crickets provides it with its primary foods.

The male seemingly lacks a flight song, instead singing persistently from elevated perching sites. He typically courts females on the ground, with tail-spreading, wing-drooping, and crest-raising displays while uttering short bursts of song. After mating, the female constructs her nest at a site ranging from ground level to as high as about ten feet above ground. Not very well concealed, these nests are often parasitized by brown-headed cowbirds. Renesting is regular following nest failure, and double-brooding has been suspected but is apparently unproven. During the period 1966–2000, the lark sparrow declined at a national average of 3.4 percent annually on breeding bird surveys, one of the highest rates reported for all grassland sparrows.

Like the plains, silky, and hispid pocket mice, the olive-backed is one of those many desert- and grassland-adapted mice that are the mammalian counterparts of the grassland sparrows. Also like the sparrows, the several species of pocket mice are often almost identical in appearance and size. Presumably only minor differences in their ecologies help to reduce competition among them. Most weigh only about a half an ounce, and all have long tails, large heads but small ears, and fur-lined external cheek pouches that can hold a substantial number of seeds. The olive-backed species is part of a group

Figure 30. Lark sparrow, male on western juniper

of pocket mice called "silky-haired," which includes the plains pocket mouse and silky pocket mouse, and these three very similar species have overlapping ranges in the central and southern plains. The pelage color of all species closely matches their environmental background, making identification especially difficult. The olive-backed is especially common in the dry northern grasslands, often with rocky outcrops, whereas the plains pocket mouse is most typical of especially sandy soils. The silky prefers areas of thin grasses of the central plains, and the considerably larger hispid pocket mouse is most abundant in the hot southwestern semidesert grasslands, but by virtue of its large body size seems relatively tolerant of cold.

Plains pocket mice dig narrow but deep tunnels in soft earth. Their tunnels may be up to about eight feet in length but only an inch in diameter, and these are kept firmly closed with plugs of earth during daylight hours. Within the tunnel system are enlarged areas where caches of as much as a pound of seeds are kept. In corn-growing areas the animals may make heavy use of it as a food source, but more generally seeds and dried fruits of native plants are the usual diet.

The short-grass plains have nearly all the same lizards and snakes that occur in the mixed-grass prairies and sage scrub, and some of them resort to spending the hot daytime hours of summer hiding in the burrows of prairie dogs or ground squirrels. Collared lizards occur in all these habitats, north from Mexico to Kansas, especially where a rocky substrate is present to provide hiding places. They are among the most attractive of all Great Plains lizards, the adults with one or two vertical black bands on an orange background, and the upper parts sprinkled with orange crossbanding and pale yellow spots on a grass-green (males) or brown (females) background. Coloration may also vary with locality and age as well as individual variation. Adult males are larger than females and have a yellow to orange throat and bluish belly markings, especially when breeding. The animals reach almost five inches as adults and are nimble runners, sometimes getting up on their hind legs and sprinting in a bipedal manner. They are often associated with rocky and canyon-rich areas, such as the Black Mesa region, but also occur widely throughout the southern plains. The collared lizard is the official "state reptile" of Oklahoma, where it occurs widely and is commonly (if inexplicably) called the "mountain boomer." There and elsewhere the animals especially favor pinyon-juniper woodlands, desert scrub, and other rather arid habitats, especially where rocky outcrops are present.

Collared lizards are insectivorous and in turn are often eaten by snakes and hawks. The males are highly territorial and perform visual displays or direct aggression toward other males. Females are less territorial but tend to be nonsocial. At least in western Texas, females emerge from hibernation in March. Mating occurs during April and May, egg-laying is done during April to June, and young appear two to three months later, during August and September. The breeding season in Kansas occurs about a month later than in west Texas, with young lizards sometimes being seen as late as October.

CHAPTER 5

Box Turtles, Blowouts, and Old Boots

The Sandhills Grasslands

Imagine a place in the Great Plains where the nights are so dark that almost every star in the visible universe can be seen, and the evenings are so quiet that coyotes can be heard yipping from miles away. Visualize a land where the nearest grocery store or filling station may be fifty miles or more distant, and the sight of a billboard sufficiently rare that one actually notices and reads it. Think of a locality where the presence of old, discarded cowboy boots stuck upside down on a fence-post may be the only sign of human influence, and where a line shown as a road on a state highway map may represent nothing more than two narrow tracks in bare sand that disappear over the far hills without so much as the slightest hint that anything or anyone might exist at the other end. It is not a land for the fainthearted, for those in a hurry to be somewhere else, or for those unwilling to feel totally alone and self-reliant. It is a land, however, of gracefully bending horizons, of waving grass and shifting late afternoon shadows, of stunning sunsets, and of inner peace. It is called the Nebraska Sandhills.

It is true that there are other sandy areas in the Great Plains, and even a Great Sand Dunes National Monument in Colorado. But the Nebraska Sandhills dwarf all these other areas—it is indeed the largest region of sand dunes in the Western Hemisphere and, unlike the Sahara Desert of Africa, is essentially fully blanketed in native grasses. Even more remarkable, if a well should be dug at the bottom of a dune, it is likely that water may be present only a few feet below. Fresh, unpolluted water may also unexpectedly flow out as an artesian spring from the base of a dune. Below this largest area of sand in North America is the continent's largest known aquifer, the Ogallala aquifer. Although the dunes may locally reach several hundred feet high, the zone of saturated sand below them may extend downward 600 feet or more from their interdune valleys, often to as deep as the sandy deposits themselves. The waters that are held by this immense reservoir are the result of downward percolation from thousands of years of rainfall. The flow of water moving slowly but inexorably eastward in water-bearing sands and gravels has its origins in montane streams and rivers along the eastern slopes of Wyoming and Colorado.

The geologic origin of these sands is still somewhat controversial, but they probably originated sometime during Pleistocene times. Immense quantities of sands and silts were then carried in by glacier-fed rivers of the eastern Rocky Mountains, such as the ancestral and necessarily much larger Niobrara and perhaps even the Platte. The sands have since been reworked by wind action into dunes, and the silts were blown southeast for varying distances, eventually to be deposited as loess up to several hundred miles away. The dunes have also been set into motion by winds of several different periods of interglacial and postglacial droughts. The current, relatively stabilized dunes are increasingly believed to be quite dynamic, their last major movements occurring perhaps as recently as 5,000 to 7,000 years ago, during a long "xerothermic" climatic period of unusually hot and dry millennia. Predominantly northwesterly winds since then have gradually shaped the dunes into long ridges and valleys that are generally oriented in a southwesterly to northeasterly direction, the ridges sometimes up to 350 feet high and often ten to twenty miles in length. It is likely that under conditions of periodic droughts these dunes have since undergone occasional losses of most vegetation, setting at least some of them into motion for decades or even centuries. Sometimes the moving dunes have thus intercepted creek valleys, forming "dune dams" that have impounded water on the upstream side, forming lakes, marshes, and wet meadows there and producing artesian springs on the lower side. At least some of the wetlands in the vicinity of the Crescent Lake and Valentine National Wildlife Refuges probably had such origins.

The botanic characteristics of the Sandhills comprise a mixture of species derived from the tallgrass and mixed-grass prairies that surround them, plus a number of specifically sand-adapted plants, especially grasses. These include many of the mixed-grass native perennials, such as little bluestem, plus several taller grasses, such as sand dropseed, sandhills bluestem (a very close relative of big bluestem), and blowout grass. Blowout grass is an especially important vegetative component, since it is able to invade areas of bare sand, or "blowouts," with its network of lateral roots and begin to restabilize the area. Other similar invaders of disturbed sand include sand lovegrass and sand muhly. Among shrubs there are the Arkansas rose and sand cherry, both of which provide important fall and winter fruits for grouse and other birds and mammals. There is also the endemic and highly endangered Hayden's or blowout penstemon, one of the few endemic plants of the Great Plains states. Other distinctive plants of the Sandhills include Great Plains yucca, also locally called soapweed, a nearly nonedible plant for most grazing animals owing to its sharp and fibrous leaves, but a useful perch for songbirds and a producer of great quantities of large and nutritious seeds that are eagerly collected and consumed by rodents such as pocket mice and kangaroo rats.

Prairie dogs and pocket gophers are now uncommon to rare in the Sandhills, perhaps because the sandy substrate makes large burrows prone to collapse, and furthermore ranchers have historically been less than friendly toward prairie dogs. However, there are many smaller rodents, especially the plains and hispid pocket mice and the Ord's kangaroo rat. Spotted ground squirrels are local and uncommon in the Sandhills, but jackrabbits as well as eastern cottontails might be encountered almost anywhere. Two harvest mice, the plains and western, occur widely in the Sandhills, as does the northern grasshopper mouse. In meadows and near water, meadow jumping mice, meadow and prairie voles, deer mice and white-footed mice, and southern bog lemmings are more likely to occur. Common predators include coyotes, long-tailed weasels, badgers, and skunks. The most conspicuous snakes are western hognose snakes and bullsnakes. Hognose snakes are toad specialists, whereas bullsnakes are generalists and can do great damage to ground-nesting birds and burrowing mammals. They can even climb rough vertical rock or cement surfaces to reach and plunder the nests of birds such as cliff swallows. By contrast, prairie rattlesnakes are rare in the heart of the Sandhills, as rocky areas are lacking. And one may often encounter ornate box turtles wandering about the dunes or down the sandy roads in a determined but seem-

Figure 31. Long-billed curlew, female

ingly mindless condition. Their persistently perambulating trait is a dangerous one, as they are often run over by cruel or thoughtless drivers or are picked up by tourists who mistakenly think they have found an attractive and free pet.

The breeding birds of the Sandhills are so different from those of the adjacent dry prairies that upon entering the area one seems to have suddenly begun a new chapter in one's ornithological memories. Suddenly horned larks begin to flush from the roadsides, their white-edged black tails providing the only easy fieldmarks as to their identity. Loggerhead shrikes periodically decorate the telephone wires that line the main roads and send occasional branches off across the dunes toward some invisible ranch. Western meadowlarks sing persistently from yuccas, and western kingbirds flutter out from the low trees that line creeks and wetlands. Sharp-tailed grouse scuttle off rapidly through the grasses, and grasshopper and vesper sparrows perch inconspicuously on low barbed-wire fences or tufts of little bluestem. Best of all, the whistles and screams of upland sandpipers and long-billed curlews proclaim that, yes, you really are back in the Sandhills, and all's right with the world.

H. Elliot McClure found horned larks and western meadowlarks to be the two most commonly observed bird species that he encountered during horseback trail surveys he made in the early 1940s, and it is unlikely that things have changed much in that regard during the six decades since. Furthermore, although the meadowlarks migrate south between November and February, the horned lark population remains fairly steady all year. Mourning dove, burrowing owls, and black-billed magpies were other relatively common species named in McClure's studies. The three most common raptors were northern harrier, American kestrel, and Swainson's hawk. Except for the burrowing owl, now rare in the Sandhills, these observations would probably still apply. Among larger mammals, the black-tailed prairie dog was most numerous, followed in turn by the black-tailed jackrabbit and white-tailed jackrabbit. Neither jackrabbit species is now common in the Sandhills, and prairie dogs are virtually absent.

It is hard to imagine a more perfect symbol of the Sandhills than the long-billed curlew (Figure 31). At least in Nebraska it is unique as a breeding species to the Sandhills. Its plumage blends well with its dead-grass background and is nearly the same creamy buff-brown color of the sand. It often stands conspicuously on dune-tops or circles about in the sky, proclaiming ownership of a wide territorial domain, just as independent-minded Sandhills ranchers are prone to do. No wonder the ranchers themselves most often choose it as their most favorite bird, and the one they consider most symbolic of their region.

Long-billed curlews arrive each spring in the Sandhills from their coastal wintering areas about the middle of April, and their glorious calls soon begin to resound across the landscape. They waste little time in getting started with their nesting. In spite of many summers of wandering about in the Sandhills, I have found only two nests—the adults are wonderfully good at distracting a person from the nest's actual location. In spite of the rather scant cover in which they typically nest, a sitting female can become virtually invisible. I once set up a blind no more than fifteen feet from a curlew nest and thought I could easily pinpoint its exact location from my peephole. Yet, each time I entered the blind I was never sure the bird was actually sitting on its eggs until I could locate its dark brown eye watching the blind intently. The bird incubated with its head held low, its long, decurved bill almost touching the ground. Typically its mate stood guard some distance away, always ready to ward off any intruders. Should it sound the alarm and take off, nearby curlews would soon also appear to join in the diving attacks that cause even an adult person to duck down instinctively.

The long bill, which is significantly longer in females than in males, would seem to be mainly adapted for probing for invisible prey such as crustaceans in wet sand rather than useful for plucking grasshoppers and other insects from the surface, which is what the birds do while on their breeding grounds. In any case, they are certainly adapted at visual foraging and probably mainly subsist on grasshoppers during their summer stay in the Sandhills. In late June I once watched a flock of more than fifty birds stuffing themselves on grasshoppers in an irrigated alfalfa field just outside the boundary of Crescent Lake National Wildlife Refuge. The group consisted of adults and just-fledged young, and within a week or two all had disappeared, presumably having begun their migration. Nebraska birds seem much more likely to travel south to the Gulf Coast of Texas than the Pacific Coast of southern California, both representing the most important wintering areas for the species. A third, apparently less important, wintering area is the vicinity of the lower Pecos River in western Texas. I have seen large groups of curlews migrating north toward the Sandhills in early April while observing lesser prairie-chickens along the upper Arkansas River of Kansas, which would be in a nearly direct line between western Texas and central Nebraska. Nationally the long-billed curlew population declined an average of 1.2 percent annually between 1966 and 2000, according to North American breeding bird surveys, and in the Central States region of the U.S. Fish and Wildlife Service (USFWS) it declined at an average rate of 3.1 percent during the same period. It is considered a species of special concern in Kansas and Oklahoma, and one recent estimate put its national population at only about 20,000 birds.

Over most of the plains states, the horned lark (Figure 32) is a widespread and fairly common breeder in various prairie habitats. In North Dakota it has even been judged to be the state's most common breed-

Figure 32. Horned lark, adult males

ing bird, and in South Dakota it ranks about fourteenth in overall frequency of breeding occurrence. Overall, it doesn't join the list of Nebraska's top twenty breeding species, but any bird survey in the Sandhills will almost certainly show it to be that region's most common breeding bird. Yet, most casual tourists are unlikely to notice the birds as they drive down Sandhills roads; their plumage so perfectly matches the color of sand that the birds are essentially impossible to

85

see before they fly, and then they offer only a few seconds of viewing time before landing and again disappearing from view.

Like many sparrows, the horned lark needs to be observed closely and carefully to be fully appreciated. Adults of both sexes are nearly identical, the female having a slightly duller face pattern and with slightly less conspicuous feathered "horns," which are only rarely raised. Unlike most songbirds, the chicks are hatched with a cream-colored down that is longest on the head and back and is soon replaced by a mottled juvenal plumage. By about two weeks, when the young leave the nest, the chicks are a curious speckled or mottled pattern that completely confused me the first time I saw one. Luckily, it was soon joined by an adult, ending the uncertainty of its identification. The white edging of the black tail is effectively hidden until the bird takes flight, making this mark useless for the identification of perched birds.

Perhaps the best feature of horned larks is the territorial song flight of the male. Like its European relative the skylark, the horned lark advertises its territory mainly by performing high, sometimes nearly stationary flights while singing loudly. Typically the bird ascends to as high as 250 feet before singing, then spreads both wings and tail and begins to glide, singing a rapid series of mostly descending notes with little real vocal patterning. If there is a wind, the bird orients into the breeze, so it hovers nearly stationary in place or may even be carried backward somewhat. This type of song is called "intermittent song." Another song type, uttered both while perched and while actively beating its wings in flight, is "recitative song." In this song type, the vocalizations may last up to a minute or more, and like intermittent song the sequence is mostly comprised of unpatterned and slurred notes. At the end of a song flight, which may have several song sequences, the bird suddenly closes its wings and drops quickly back toward earth, opening its wings and braking only when dangerously close to the ground. It then lands and may start singing again from a conspicuous location.

Horned lark territories are relatively large, averaging about four acres, and include all the resources needed to provide for successful breeding. The birds appear to be completely monogamous, but they may change mates between successive breeding seasons. However, should the males from an adjoining territory die, a neighboring male may expand his territory to encompass that of the other pair and take on the additional female as a second mate. During the period 1966–2000, the horned lark declined at a national average rate of 1.95 percent annually on breeding bird surveys, one of the higher rates of decline among grassland birds.

After young horned larks leave their nests, they follow their parents about on Sandhills dunes, where their salt-and-pepper plumages make them difficult to see by avian predators such as loggerhead shrikes, magpies, and crows. Loggerhead shrikes are presumably named for their relatively large heads, which are associated with massive beaks and powerful jaw muscles used in subduing and killing small prey. The origin and meaning of "shrike" is less certain. The genus *Lanius* is Latin for "butcher," and "butcher-bird" is a fairly common vernacular name for the birds too. Like butchers, shrikes may impale their recently killed prey on a tree thorn or on the barb of a barbed-wire fence, leaving it there temporarily until it is eaten. This curious trait is shared by no other group of birds and would seem to be of limited survival value except in situations where the prey is only periodically superabundant.

Shrikes are easily the most raptorial of all North American songbirds, and wherever they occur around the world they are noted for their killing abilities. Their prey not only includes such expected ones as large grasshoppers and crickets but also extends to very small rodents and many small songbirds, especially nestlings. Adult shrikes average less than two ounces in weight and yet may take on fairly large prey. One early study quoted by A. C. Bent (1950) indicated that among eighty-eight stomach samples, 72 percent of the identified items were insects or spiders, and 28 percent were vertebrates. During winter, when insects become unavailable, the incidence of vertebrate prey sharply increases. Birds as large as adult chipping sparrows, house sparrows, and myrtle warblers have all been recorded as prey; the average weights of these species may be more than half that of their killers. Mammals as large as thirteen-lined ground squirrels may be taken; adults of these animals would easily outweigh the shrikes. Among sixty-four vertebrate prey reported from Oklahoma, twenty-four were birds, twenty were reptiles, and twenty were mammals.

Loggerhead shrikes are common nesters in the Sandhills. They seem to favor low, dense trees for nesting, such as Russian olives, and because of these locations it is usually possible to peer easily into the nests. Generally well concealed but bulky structures, the nests typically

Figure 33. Northern grasshopper mouse, calling (above) and relaxed (below) postures

have four or five eggs. Unlike hawks and owls, the female begins incubation with the last or next-to-last egg, so that all the chicks tend to hatch at about the same time. However, there is often a clear hierarchy of strength among the chicks, so that starvation of some weaker chicks is likely to occur in favor of the stronger ones. Although in some regions two broods might be produced in a single breeding season, one seems to be more typical, with renesting likely following an early failure of eggs or young. Nationally, the population of loggerhead shrikes has been declining at an annual rate of 3.6 percent, one of the highest rates of all grassland birds.

Mice of almost any species are common prey for shrikes, especially those likely to be active during daylight hours, such as the grasshopper mouse (Figure 33). One of the first grasshopper mice I saw in the Sandhills was on a family camping trip, when my younger son spotted one, and decided to catch it and keep it as a pet. Having run it down and caught it, he had no place to put it except his pocket. He did so but soon let out a loud yelp after it fiercely bit him. I should have warned him that grasshopper mice are the most carnivorous of our native mice and will not hesitate to attack another animal. He then regretfully decided it was best to release the mouse, which was a beautiful cinnamon-yellow adult.

Grasshopper mice are certainly among the most attractive of our native mice; in addition to the cinnamon pelage morph just mentioned, they also have a gray version, thus being the only Great Plains mouse occurring in two quite distinct color types. It is also unusual among Great Plains mice in having a quite short tail and a distinctly "stocky" body, powerful jaw muscles, and long fingers and claws that are used in grasping its prey, mainly large grasshoppers. Although small mammals may at times also be attacked, invertebrates evidently constitute their primary diet.

As active predators, grasshopper mice have evolved traits parallel with those of some larger predators including large, defended, and scent-marked territories up to an acre or more in size, complex social behavior and courtship, and long-distance communication. Like coyotes or wolves, this consists of a prolonged howl, uttered with the head stretched upward and while standing on their two hind legs. It is a high-pitched, pure-tone vocalization (of about 12,000 cps, or about the same pitch as kinglets), lasting less than a second, but capable of carrying a considerable distance. Males primarily call, especially when seeking females, but females may rarely also utter the same or a similar call. Larger animals have deeper voices than smaller ones, and so age and body-size information easily might be transmitted. Like coyote or wolf howls, it probably conveys detailed information as to the caller's species, sex, and individual identity, as well as its location. It perhaps provides additional information as well, such as relative social or reproductive status. Social dominance in these tiny predators may be just as important as it is in the large canids.

Grasshopper mice are also unusual in that the male remains close to his mate following birth and may help defend their young. He also may help teach the youngsters the difficult art of capturing live prey. Likewise, females have unusually strong maternal bonds and care for their young for several weeks after birth. Although relatively mobile, grasshopper mice sometimes use the burrows of prairie dogs, kangaroo rats, or pocket mice for nests or for caching food.

Pocket mice are even more common in the Sandhills grasslands than are grasshopper mice. They have all the charm of a Walt Disney mouse; their bulging eyes, large head, long tail, and tiny forelegs make them appear more like something out of a cartoon than a real animal. Like kangaroo rats their favorite habitats are sandy ones in dry habitats, where they can easily dig, and where seeds can be collected and kept for long periods without danger of becoming wet and moldy. They make nightly excursions out of their burrows to find small seeds, which they stuff into their fur-lined external cheek "pockets" as rapidly as possible, then hightail it back to their burrows to deposit their treasures in seed caches. Like kangaroo rats, their hind legs are relatively long, so their common mode of locomotion is by hopping, at least when they are in a hurry. The long tail probably provides a useful counterbalance for such locomotion. By day, the naturalist will only find a few tiny footprints to mark their passing. Or, by capturing the mice and dusting them with a nontoxic fluorescent powder, the trails made by each marked mouse can be easily tracked at night with the aid of an ultraviolet lamp.

Plains pocket mice are extremely common in the Sandhills; they favor a combination of sandy substrates and patches of open ground. In one study, of 770 rodents captured on Arapahoe Prairie in the western Nebraska Sandhills, the plains pocket mouse comprised 68 percent; the hispid pocket mouse, 3 percent; and the Ord's kangaroo rat, 2 percent. Several other rodent species made up the remainder, with grasshopper mice well represented at 12 percent. The other two Nebraska pocket mice are more prevalent in the western high plains steppes having a combination of gravelly or sandy to clay soils, rocky outcrops, short grassy or sage vegetation, and considerable bare ground (olive-backed), or in western arid prairies with sandy or loamy soils and limited bare ground, interspersed with sparse grasses, weeds, and shrubs (silky).

The burrows made by plains pocket mice are inconspicuous; their openings are only about an inch in diameter and are kept plugged with sandy soil whenever the occupant is within. Each burrow may be up to about seven to eight feet in length, with periodic widening for seed storage. They typically have only a single opening. The seed caches probably allow the animals to remain underground for prolonged periods when necessary, and little or no surface activities occur during the coldest months of the year. Like other pocket mice, the animals don't need to locate freestanding water; instead metabolic water is produced through digestion.

Foods of the plains pocket mouse are almost entirely of seeds, but when seeds are scarce, arthropods such as ants may be eaten. The animals are quite catholic in their tastes and won't hesitate to collect corn, should they happen to be living close to a supply of such foods. They are also good climbers and will often climb to the top of a plant such as a sunflower to collect its seeds directly. The pocket mice of the Sandhills region have a pelage color that closely matches their yellowish sandy substrate; in other areas where the sand is of a different color, the pelage color of the local population typically adapts to match it too.

If the pocket mice of the Sandhills are considered "cute," then the Ord's kangaroo rat (Figure 34) goes one step beyond in achieving a maximum of babylike human features within a rodent's morphology. Their eyes are larger, their faces are more rounded, and their foreheads are more bulging. Their hands are relatively much smaller than their hindlegs, so they are more likely to stand erect in humanoid fashion, especially when holding food or stuffing it into their capacious external cheek pouches. Because they are adapted to a desertlike climate, they require no freestanding water, and they produce a minimal quantity of urine. As a result, they are preadapted to make extremely attractive pets, and during my early years at our biological station, it was a common practice to go out at night with an insect net and a flashlight in search of "roo-rats" to capture as pets. Later, fear of Hanta disease and other possible dangers gradually put this activity to a halt, but I know of students who kept their animals alive for several years. Captives are known to have survived for as long as seven years.

The lives of kangaroo rats are similar to those of pocket mice. Because of their larger size and greater food requirements, they have larger home ranges, of up to about three acres. Their density in very favorable habitats such as the Sandhills may reach up to twenty animals per acre. They are almost entirely nocturnal, as suggested by their huge eyes, and can easily hop for distances of as much as five feet, especially when frightened. They can also hop vertically, or even backward, when trying to escape a predator. The inner ear chambers of kangaroo rats are extraordinarily large, which provides them with an acute sense of hearing, especially for low-frequency sounds such as those that might be made by an approaching predator.

Figure 34. Ord's kangaroo rat, adult

Unlike grasshopper mice, pair-bonding in kangaroo rats and pocket mice is lacking, and except for brief sexual contacts the animals live solitary lives. Sexual behavior, including copulation, probably occurs in the female's burrow rather than on the surface. It is likely that males can detect the odor of a sexually receptive female, and can track her back to her burrow. Life spans in the wild are thought to average about one year; owls, snakes, and weasels are likely important predators.

Pocket mice and kangaroo rats have little protection from predators such as shrikes and hawks other than quick reflexes, bounding jumps, and nervous dispositions. They sometimes may also become trapped in their burrows by snakes such as bullsnakes or hognose snakes. The western hognose snake is certainly not rare in the Sandhills, and it finds its preferred habitats in sandy areas where it can eas-

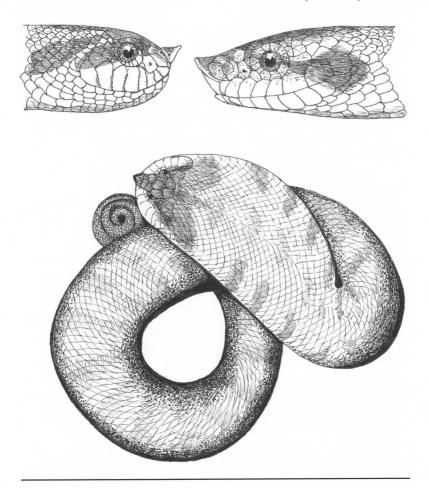

Figure 35. Western (top left) and eastern (top right) hognose snakes. The defensive neck-widening posture of the eastern hognose snake is also shown (below).

ily burrow in order to maintain an ideal body temperature. From Kansas south, it is joined by the eastern hognose snake, which also favors sandy areas. All hognose snakes have curiously upturned snouts, which aid in their digging ability (Figure 35). The western hognose also has a much more obviously upturned "nose" than the eastern, making it fairly easy to recognize. Much of their food is obtained by digging, during which they might expose hidden toads, the eggs of reptiles, as well as small snakes and lizards. Salamanders and frogs are sometimes also eaten. Although not dangerous, hognose snakes have two enlarged teeth at the rear of the upper jaw, which help in "deflating" toads that have become enlarged with swallowed air as a defensive strategy. They also secrete fluids from their adrenal glands that quickly detoxify the powerful poisons produced by toads.

The defensive behavior of hognose snakes is especially interesting. Hissing is common, as is striking toward the intruder, with the neck area expanded laterally to form a cobralike hood. If these tactics fail, the snake is likely to roll over and "play dead," usually first disgorging any recently eaten food.

Hognose snakes are diurnal, emerging from winter burrows in spring and typically mating in April or May. Eggs are laid in June or July, with clutch sizes varying greatly. From as few as about four to as many as sixty-one eggs may constitute a clutch, the eastern hognose averaging more than the western in this regard. Incubation requires about two months. The western hognose snake reaches a maximum length of about thirty-five inches as compared with a maximum of forty-five inches in the eastern species, whose range overlaps the western to some degree. In both species, females grow larger than males, which have relatively longer tails than females.

Ornate box turtles (Figure 36) are easily the most approachable of our Sandhills wildlife. During the months of June and July especially, when males are on the lookout for receptive females, they are often quite literally "on the road." Males are easily identified by their bright red rather than brownish-yellow eyes, and they average slightly smaller than females as adults. Both sexes have an attractive pattern of yellowish spots and vertical streaks coursing down their dorsal shell, or carapace, looking as if they might have fallen asleep under a housepainter who was careless about where his paint had dripped.

Their appealing color pattern, plus the ability of the turtle to fold its lower shell, or plastron, up so that head, tail, and feet are all tightly enclosed in a rounded "box," adds to the charm of the animals. This adaptation helps protect them from attacks by coyotes and raccoons, but it is an inadequate defense against cars and trucks. It seems that box turtles are burdened with an irresistible urge to always be someplace else, and animals of either sex might travel up to 300 feet in a day and have home ranges of up to five acres. Not only are males often moving about searching for females, but pregnant females also may travel great distances in search of a perfect place to lay their eggs.

Figure 36. Ornate box turtle (above), tiger salamander (middle), and Plains spadefoot (below). The eggs and larvae of the salamander (two larval stages) and spadefoot tadpole are also shown, as well as the "spade" of the spadefoot (as seen from below).

Because of this tendency, humans often encounter the animals while driving down Sandhills roads, and all too frequently run over them purposefully. In one Kansas study, fifty-three box turtles were counted during a two-week period in Cheyenne County, of which nearly half had been killed by vehicles.

In Nebraska, in addition to road-kill losses, vast numbers of ornate box turtles are picked up every year by animal dealers, to be shipped out of the state and sold in pet stores. The animals rarely live long in captivity, for most people assume they will live on a vegetarian diet of lettuce and carrots, when in fact they are omnivores and need a substantial amount of protein in their diet. Until 2001, when the laws were finally tightened, the Nebraska Game and Parks Commission refused to protect these helpless animals from such exploitation. Between 1994 and 2000, more than 58,000 ornate box turtles were reported to the commission as having been sold and exported from Nebraska, and this is only one of thirty species of Nebraska's herps that were thus exploited. The total numbers of all herps reportedly sold and exported during that period approached 300,000 animals. Since only about a third of the permit holders bothered to submit such activity reports, it is likely that the total take was actually much higher.

It seems strange to find tiger salamanders (Figure 36) in the dry Nebraska Sandhills, but they are actually quite common there and even extend out into the more arid short-grass plains, where they often occupy prairie dog burrows to avoid desiccation in the summer heat. The Sandhills wetlands provide fine habitat for them and may help put them out of easy reach of many predators that are common where there are more solid substrates. The tiger salamander has a beautiful overall pattern of orange-yellow to dusky yellow or brownish-olive spots and bars on a blackish to olive-green background. Most of these color and pattern variations are associated with different subspecies; a "barred" race of the tiger salamander is found in the plains states south of the Niobrara River; other races with differing pigment patterns occur to the north and east. Tiger salamanders eat a wide variety of foods but are especially fond of earthworms. One would imagine that such a bright color pattern might warn possible predators of a distasteful if not poisonous meal, but this is not the case, and the animals are eaten by predators that include great horned owls and hognose snakes. Breeding occurs in water, the young having a prolonged larval period similar to those of toads and frogs, but the tadpolelike larvae retain external gills for much of their growth, and in some entirely aquatic forms the adults may retain gills for their entire lives.

The plains spadefoot (Figure 36), an unusual species of toad having vertical rather than horizontal pupils, also likes sandy soil, and with the aid of its spurs on the hind feet it is an effective digger. At times it may burrow as much as two feet underground, or until it encounters moist soil. Like the tiger salamander, it moves too slowly to avoid capture by snakes and other fast-moving predators. On the other hand, it has a skin rich in secretions that can produce severe allergic reactions when handled or even asthma attacks in humans. Its primary foods are small insects.

Both species winter below the surface and emerge with warmer weather and spring rains. Male spadefoots then gather around pools and call loudly, usually in grouped choruses. On the other hand, tiger salamanders court in water, with mutual body-rubbing a significant part of their breeding activity. The spadefoot female lays her eggs in small clumps of up to about 250 individual eggs and in total may produce as many as 2,000 eggs in a single cycle. The tiger salamander female lays her eggs singly or in clumps of two or three, and as many as 1,000 are produced in a single cycle. No parental care is given by either species. Salamanders require two years to become sexually mature, remaining as gilled larvae their first year. In some areas, where environmental conditions are harsh, they may even retain their gills into adulthood.

CHAPTER 6

Rattlesnakes, Roadrunners, and Rock Wrens

The Arid Shrubsteppes

The sandsage grasslands community type of the southwestern Great Plains occurs locally as far north as extreme southwestern Nebraska. Yet, it mainly extends from the sandy edges of the Arkansas and Cimarron valleys of southwestern Kansas and adjoining Colorado south through the rolling plains of western Oklahoma and the northern panhandle of Texas to the Staked Plains arid semidesert grasslands along the border of northwestern Texas and southeastern New Mexico.

Sandsage, like its many close western relatives, is a drought-adapted, long-lived shrub that grows up to about three feet high. Its leaves are small and filament-like, but are rich in proteins. The leaves of sandsage and the acorns of a dwarf, sand-adapted oak, the shinnery oak, provide nutritious all-season foods for a variety of wild birds and mammals, and both of these brushy plants offer summer cover for nesting and brood-rearing for lesser prairie-chickens, scaled quail, turkeys, and other birds. The current remaining but still declining range of the lesser prairie-chicken can be closely correlated with the distribution and abundance of these two native plants. From western Oklahoma south, greater roadrunners also are regular residents of sandsage.

The sandsage and shinnery habitats also provide hiding or escape cover for a wide variety of small mammals and singing or lookout posts for several native sparrows such as Brewer's and Cassin's sparrows. Among the typical small mammals of shinnery oak are the plains, silky, and hispid pocket mice, Ord's kangaroo rat, spotted and Mexican ground squirrels, northern grasshopper mouse, and western harvest mouse. Larger mammals include the black-tailed jackrabbit and eastern cottontail. The western box turtle, about twenty-five species of snakes, and eleven species of lizards also occur in shinnery oak communities near their southern boundary in eastern New Mexico and adjacent Texas, but none seems to be confined to it. The range of the Kansas glossy snake most closely corresponds to that of the sandsage-shinnery community type and extends north from Texas to southern Kansas.

Farther north, in extreme western Nebraska, western South Dakota, and southwestern North Dakota, another type of sage-permeated brushland occurs and extends westward across the arid western plains almost to the Pacific Coast. This is the community type dominated by big sagebrush and some lesser relatives, such as dwarf sage and long-leaved sage. Big sagebrush has one of the largest distributions of all North American shrubs, once covering roughly 150 million acres throughout the arid west. Within its vast domain, many species of birds, mammals, and reptiles have specifically adapted to its presence, including the sage sparrow, sage-grouse, sage thrasher, sagebrush vole, pygmy rabbit, and sagebrush lizard. The green-tailed towhee, gray flycatcher, and vesper sparrow are also very common in sage habitats but are not so closely tied to them.

In sage-dominated communities, the soils tend to be hard and rocky rather than sandy and permeable, making it difficult for digging mammals to excavate their burrows. Often a calcareous hardpan layer develops at or not far below the surface. In such "caprock" areas, excessive mineral salts place limits on root penetration and water availability for most plants. Water that temporarily accumulates here through rainfall often becomes too alkaline for some animals to drink and tends to evaporate quickly. Many of the most typical land vertebrates manufacture metabolic water rather than depending on surface moisture supplies. Here too only those plants tolerant of highly alkaline conditions are most able to survive; some of these plants absorb and store toxic minerals such as selenium, turning them into "locoweeds." Green photosynthetic leaves are valuable items to these plants, and they cannot readily afford to give them up to grazing animals. Spiny leaves or stems and vile-tasting plant juices rich in poisonous alkaloids also help protect arid-land plants from undesirable attacks by herbivores.

Over much of the region now associated with big sage, eroded canyon and bluff topography is typical. Glaciers never reached these isolated places, and the work of wind and water over countless millennia has sculpted the landscape into eroded "badlands" of all types. Here bluffs or clifflike promontories above valleys or tablelands provide perches and nest sites for raptors such as ferruginous hawks, golden eagles, and prairie falcons, as well as nesting crevices for smaller birds such as rock and canyon wrens, cliff and violet-green swallows, and white-throated swifts. Here too bobcats regularly prowl, and prairie rattlesnakes seek out rodents on warm summer nights. The persistent calls of common poorwills echo through the canyons, as do the occasional notes of great horned owls. It is a time when sitting by a small campfire and hearing these nighttime sounds provides a special sense of contentment and of belonging to the land.

In the early spring of 2001, I was once again feeling my recurrent need to watch the dawn courtship displays of prairie-chickens, an annual ritual of mine going back some forty years, or ever since I moved to Nebraska. However, after a controversial hunting season in eastern Nebraska, the population of greater prairie-chickens frequenting my favorite lek near Burchard Lake in southeastern Nebraska had dropped to a discouraging all-time low of only four males. Therefore, I decided to head toward southwestern Kansas in order to observe and study the displays of its nearest relative, the lesser prairie-chicken. I had never seen lesser prairie-chickens courting on their leks and indeed had only seen a few of these birds at all. Although once occurring in the sandy grasslands of southwestern Nebraska, lesser prairie-chickens have since retreated into a still-diminishing core of their original range, the sandsage and shinnery oak brushlands. In the northern part of this region, only the sandsage occurs, but where it is supplemented southwardly by shinnery oak, the population of lesser prairie-chickens is generally higher.

Since mid-April is the peak period for display among greater prairie-chickens in southern Nebraska, I decided that a week or so earlier in April should be perfect for Kansas birds. As I drove west on I-80 and followed the Platte River, I could see a few remaining flocks of sandhill cranes still feeding in the nearby fields, but already the fields were being punctuated with territorial male red-winged blackbirds and a few scattered western meadowlarks. Then I turned south, past Harlan County Dam, with its slowly gyrating flocks of American white pelicans and its dark chevrons of migrating double-crested cormorants, and entered central Kansas.

I was headed toward Garden City, the nearest area where lesser prairie-chickens were still to be found in good numbers. Here the Arkansas River cuts a lazy, undulating path through western Kansas, strewing sand deposits along either side of its course. Sandsage shares these sites with a variety of native grasses and occasional yucca plants. In more open areas and slightly heavier soils, black-tailed prairie dogs and burrowing owls still eke out hardscrabble lives. There, ferruginous hawks make their perilous living off the rapidly declining populations of prairie dogs.

A small group of us set out late in the afternoon for a lek about ten miles southwest of Garden City, where I set up my blind on a sandy slope, and where eighteen to twenty male lesser prairie-chickens had regularly displayed. Indeed, we flushed that many as we approached, and I felt confident that the next morning would be successful.

There was a full moon that night, so we arrived early at the lek about 6:00 AM, almost an hour before sunrise. The males were already present on the ground, calling in a way that I might not have even recognized as coming from prairie-chickens if I hadn't already heard recordings of their calls. As darkness gave way to dawn, it was evident that nearly twenty males were present, just as I had hoped. They paid almost no attention to the blind, except when camera sounds startled them. Their performance struck me as something resembling the choreographed drama of the greater prairie-chicken, but this was being performed in an entirely distinctive manner and on a very different ecological stage. The birds' movements were surprisingly fast, and their aggressive cacklings were unusually high-pitched. The repeated threats made by the males at their territorial boundaries were apparently mostly bluffs, as in contrast to greater prairie-chickens I never saw an actual fight.

Between such ritualized threats the males cocked their tails, raised their paired and long neck pinnae, inflated their reddish "air sacs," and expanded their bright yellow skin "combs" over their eyes. They then uttered an oddly melodious sound somewhat similar to that made by air bubbling to the surface of a thick broth. They quickly flashed their tails open and shut at the end of each display sequence. Rather than the loud, dovelike "booming" of the greater prairie-chicken, this display vocalization is sometimes called "bubbling" or "tooting." It does not seem to carry as far as the greater's "booming," which under ideal conditions can easily be heard more than a mile away. Yet, the collective noises of the males produced a somewhat melodic and nearly continuous sound background, punctuated by frequent cackling and sometimes also querulous whining notes. Often two males called in synchronized concert, the birds occasionally standing only a foot or so apart, a degree of spacing tolerance that I have never observed in greater prairie-chickens.

As the sun gradually rose, a few females slowly made their way through the group of males. These females seemingly paid no attention to the individual males and gave them not the slightest degree of apparent encouragement. Finally, one paused and squatted long enough to allow a male to mount her, thus completing the drama.

The reproductively most successful male of any prairie grouse lek is that individual old enough, experienced enough, and strong enough to establish dominance over every other male in the lek. Thereby this single alpha-level male can establish a centrally located territory, one that other males dare not enter should a female be attracted to him. When a female is soliciting in the dominant male's territory, no other male dares challenge him. Such "master cocks" also must be virile enough to fertilize most if not all the females coming to visit the lek, sometimes as many as several females in a single morning. Here "survival of the fittest" simply means mating of the fittest, in one of the clearest examples of Darwinian sexual selection to be seen among all North American birds.

If the lesser prairie-chicken is the trademark bird of the sandsage, then the sage-grouse is the counterpart species of big sagebrush. Like the lesser prairie-chicken, the distribution of the sage-grouse is correlated almost exactly with that of a sage species, big sagebrush, and this plant plus its very near relatives provide an even higher percentage of the sage-grouse's year-round food requirements. The range of the sage-grouse once approximated that of big sagebrush, or nearly 150 thousand square miles, but in recent decades the distributions of both have become greatly retracted, and sage-grouse now enter the Great Plains states only in extreme southwestern North Dakota and western South Dakota.

Recently, ornithologists have discovered that, instead of a single species of sage-grouse, there are actually two coexisting species, so that the one occurring in the Great Plains region has been renamed the greater sage grouse, while the other has been called the Gunnison sage-grouse, reflecting its limited distribution in the Gunnison River basin of Colorado. The greater sage-grouse has an especially appropriate name inasmuch as adult males weigh on average about seven pounds, or about four times as much as adult male lesser prairie-chickens, making them by far the largest of all North American grouse. Only predators such as coyotes, golden eagles, and bobcats are likely to pose significant threats to these powerful birds.

In their social behavior, greater sage-grouse somewhat resemble lesser prairie-chickens in that a lek-based breeding system exists. However, in the greater sage-grouse lek, sizes of healthy populations may average fifty or more males rather than the ten to twenty birds more typical of lesser prairie-chickens. With the greater competition among males in such large groups, it should not be surprising that serious fights are more frequent and more intense, and the overall effects of sexual selection are more clearly apparent in sage-grouse. Not only are the males extremely large, but they average almost twice the weight of females as compared with a nearly even adult weight ratio in lesser prairie-chickens. Furthermore, the genetic payoff for alpha males is greater in sage-grouse than in prairie-chickens, for a dominant male sage-grouse might easily fertilize nearly all the females in a lek of fifty or more birds during a single spring. Thereby he is passing on his own specific genes over much of the local population in the next generation, assuring that genes for large male size, virility, and social dominance will continue to be perpetuated.

Golden eagles (Figure 37) are by no means limited to the sage-steppe and rimrock topography of the American West; they also occur wildly, as for example over the steppes of Central Asia. There and elsewhere vast expanses of open country and abundant mammalian prey, especially large hares, provide a food base. The first nest of a golden eagle that I ever climbed to was on a mountain ledge overlooking a vast area of wetland tundra in the Kuskokwim River delta of western Alaska, only a mile or so from the edge of the Bering Sea. It was mid-June, and two well-grown young were in the nest. The floor of the nest was littered with the remains of arctic hares and some feathers of willow ptarmigans. From it, nothing could be seen but flat tundra to the

103

Figure 37. Golden eagle, adult

very horizon toward the east and south, the greatest waterfowl breeding area in all of North America. A similar nest in western Nebraska or the western Dakotas would probably have had the remains of jackrabbits and perhaps sharp-tailed grouse. Thus, the golden eagle is a widespread, highly adaptable bird, the top predator among North American raptors, but one that is especially associated with the presence of jackrabbits and hares over most of western North America.

In western Nebraska and Wyoming, where I have seen many nests, rimrock outcrops with near-vertical cliff faces having a few inaccessible ledges near their top seem to be favored sites. For many years, a nest was situated on the dramatic north face of Jail Rock, a well-known and nearly unscalable landmark along the Oregon Trail in western Nebraska. The birds there were highly visible but seemingly safe from every peril, until an unknown person with a high-powered rifle but a low-powered brain killed the female on the nest. The remains of the nest could still be seen more than twenty years after it was last used, and perhaps someday it will be occupied again.

Golden eagles have enormous home ranges, of up to about 100 square miles, within which no jackrabbit can ever feel secure. Like buteo hawks, the birds may circle silently on their broad wings for hours on end, their keen eyes evidently able to detect prey from heights of 1,000 feet or more, then plunge down in a dive almost as impressively steep as that of a peregrine falcon, and with equally deadly results. Their two-inch-long rear talons can penetrate to the vital organs of even quite large mammals, and a large female weighing ten pounds or more will not hesitate to take on a six-pound mammal such as a large jackrabbit or arctic hare.

If the golden eagle is thought of as regal enough to be used on the flags and symbols of many countries, the greater roadrunner (Figure 38) would never be nominated for representing anything that serious. It is very difficult to separate the actual living roadrunner from its cartoon equivalent, for every American raised on television during the past several decades can tell you all about the lifelong enmity between roadrunners and coyotes, even if they might be unable to identify a living representative of either. Yet, greater roadrunners are at least as fascinating as their cartoon version, although the birds rarely if ever run full speed down the middle of roads. They are, however, great runners, an especially useful trait for chasing and catching small lizards or snakes. Attempts to measure the species' top running speed

Figure 38. Greater roadrunner, adults

have resulted in estimates of up to about fifteen miles per hour for distances of about 300 yards. The Oklahoma ornithologist George Sutton once judged that the birds can reach eighteen miles per hour, or about the average maximum top running speed of an adult human, at least for short periods.

Roadrunners have several other unusual traits, these at least in part reflecting the fact that they are actually large, terrestrial cuckoos. Their strange toe arrangement, with two toes pointed forward and two backward, betray their cuckoo ancestry. What if any advantage this odd toe arrangement might provide in running is debatable, but this forward-backward toeprint has added to their mythic character. They also "sing" in a strange, cuckoolike manner, starting their repetitive cooing notes with their beak pointed vertically downward, but gradually raising their head with each note until it is pointed upward as the pitch of the individual notes descends. The birds have been found to consume almost every kind of animal they can catch, their prey ranging from small insects, spiders, and scorpions to rodents, nestling birds, bird eggs, and any small ground-living birds they are able to capture unaware. George Sutton once watched two captive roadrunners fighting over a cotton rat they had collectively managed to kill after a prolonged struggle. In true Solomonic fashion, Sutton finally cut the rat in two and let each have half (Bent 1940).

Roadrunners are members of the cuckoo family and as such have distinctive, evocative voices. Likewise, all the members of the nightjar family have similarly powerful abilities to stir the imagination with their songs as well as their abilities to remain essentially invisible in broad daylight. The common poorwill is no exception to this statement. I heard them singing in rocky canyons during all of the seventeen summers that I taught at our field station in western Nebraska, but I never actually observed a live bird there, save for a few that momentarily appeared, fluttering up mothlike in the glare of my car's headlights when I drove up dusty roads during ink-dark nights.

Indeed, the only poorwills I have seen well were a few that have inexplicably turned up around Lincoln during migration. One of these was picked up by our local wildlife rescue group, and the person assigned to look after it invited me to come and photograph it. When I got there, it resembled nothing more than a random pile of lint and feathers, for the person caring for it had unfortunately not noticed that it had moved and was resting on the living-room rug and, in stepping on it, had caused all its tail feathers to be pulled out. The bird was not suitable for photographing, but I still marveled at its disruptive feather patterning. Additionally, when approached closely, it would suddenly open its enormous gape and make a snakelike hissing sound that had a distinctly startling effect. Because of their specialized food requirements, consisting of live insects that are normally caught on the wing, poorwills and their relatives such as nighthawks and whip-poor-wills are extremely difficult to keep alive in captivity, and this one did not survive long.

Not only are adult poorwills and other nightjars difficult to see when they are perched on the ground, but their nests are almost impossible to locate. I have personally stumbled across two or three nighthawk nests, but know of only a few nest records of common poorwills for all the Great Plains states. Only three confirmed nests in Kansas were found during the six years and nearly 12,000 hours of that state's breeding bird atlas project. Nests in Kansas reportedly have typically been made on bare patches of gravel or on low, flat rocks and usually placed close to a clump of grass or weeds, providing some shade from the sun. However, one nest was found on bare ground in an alley in Manhattan. Instead of being spotted and concealingly colored, the eggs are nearly pure white. The nestlings are initially downy and almost uniformly buff-colored, but soon acquire the concealing pattern of the adults.

It is now known that at least some common poorwills hibernate in rock crevices during the colder months, a trait otherwise virtually unknown among North American birds. The first individual found in this remarkable condition was determined to use the same resting niche every winter for four successive winters. Very few such torpid birds have ever been found in the wild, but captive birds can be induced to enter a state of torpidity. In one case it was determined that a ten-gram fat deposit would support a torpid poorwill for 100 days.

Like roadrunners and nightjars, Texas horned lizards (Figure 39) are the stuff of legend, in which truth often is hard to separate from fiction. Stories of the lizard shooting blood out of its eyes are at least part of the animal's strange folklore, and in this case actually are truth rather than myth. When disturbed, the lizard's head develops a higher temperature than the rest of its body, and the slightest pressure will cause blood vessels around the eyes to burst, producing a small stream of blood that may at times be squirted forward. It is thought that this behavior might serve as a distracting or defense behavior, but it has also been observed that such behavior has not prevented a coyote from consuming the animal after first licking away the blood. However, its generally spiny integument may have some protective function and certainly adds to its grotesque appearance. Regrettably, this animal's strange appearance also makes it a visually appealing pet, but horned lizards are very difficult to keep alive in captivity, and such attempts should be discouraged. Additionally, the horned lizard is apparently declining over much or all of its range.

Horned lizards are active from about April to September in the cen-

Figure 39. Texas horned lizard (above), defensive blood-spraying toward swift fox (middle; adapted from a photo in *Copeia* 1992:519), and precopulatory and copulation behavior (below; after sketches in *Herpetologica* 37:136, female stippled).

tral and southern Great Plains. Breeding in Kansas probably is centered during May and June, as also has been reported for Texas. Average clutches are twenty-three to thirty eggs, with females sometimes producing multiple clutches during a single season. Hatchlings appear from June to September, emerging from subterranean nest chambers. Two years are required for attainment of sexual maturity, and although slow-moving, the adults may have home ranges of as much as about 2.5 acres. They have survived up to almost ten years in captivity.

In essentially the same habitats as are preferred by common poor-wills and horned lizards, rock wrens are likely to be found. I remember finding the nest of a rock wren at the foot of Courthouse Rock, a large and historic monolith located adjacent to Jail Rock and near the North Platte River of western Nebraska. The nest was placed in a seemingly natural crevice amid fallen rocks, its entrance marked by a distinctive pavement of small pebbles located around the opening in the rocks. Although reminiscent of the bits of dried cow dung that often mark the entrance of a burrowing owl's nest burrow, these pebble accumulations are of less certain function. Perhaps they help make the entrance somewhat smaller and harder for rodents or other animals to enter, or possibly they provide a landmark in helping the adults locate the nest among the rubble. It is unlikely that they simply accumulate as the inner nest chamber is cleared of debris, since some materials are clearly brought in, and in one remarkable case mentioned in Bent's life histories (1940), over 1,600 items were present, including almost 500 small stones of very uniform size.

During the years I taught there, rock wrens were also common in a canyon leading up from our biological station to the main paved highway linking Ogallala and Kingsley Dam. Walking up this steep canyon on a hot summer day was a certain way to generate loud groans and complaints from ornithology students, but also a certain way to see and hear rock wrens, the even louder inhabitants of such locations. Like house wrens, rock wrens are prone to sing all day long, even after nearly all other birds have become quiet and inconspicuous. They seem oblivious to any dangers such visibility might generate, apparently operating under the strategy that such tiny prey are not worth chasing for most predators.

Rock wrens are only slightly larger than house wrens, adults averaging about fifteen versus ten grams, but their vocal volume is substantially greater. Like all wrens, they are highly territorial and more than make up for their dull plumages with their loud and persistent singing. Their songs are also more grating, more mechanical, and less musical than those of house wrens. In some barren, vegetation-free, and waterless landscapes such as the badlands of South Dakota and western Nebraska, they might well be the only bird species to be seen or heard. At such times, and in such places, their presence is especially welcome. During the period 1966–2000, the rock wren declined at a national average rate of 1.7 percent annually on breeding bird surveys.

When clambering about rock faces in search of rock wren nests or other rock-inhabiting wildlife, it is always well to be very careful about not putting one's hands in places that can't be seen clearly. There are few sounds more disconcerting than the sudden sound of a rattlesnake coming from some very nearby but uncertain location. It is a sound that must be equally frightening to many other species, as rattlesnake mimics include not only some nonvenomous snakes that shake their tails vigorously in defense but also burrowing owls trapped in a burrow that utter similar rattling notes. Some insects or their pupae produce surprisingly similar rattling sounds when they are disturbed, and various small rodents also produce sounds similar to that of rattlesnakes.

Once one realizes that rattlesnakes produce their rattles as a means of trying to avoid conflict and thus not wasting their venom rather than as an indication of intended attack, it becomes somewhat easier to appreciate rattlesnakes on their own terms. Prairie rattlesnake bites are certainly not to be taken lightly, but they are appreciably less lethal than the bites of timber rattlesnakes, for example. Furthermore, only about a dozen deaths per year are caused by all venomous snakes in the entire United States, or roughly one-tenth of those caused by lightning strikes. Two students were bitten by prairie rattlesnakes during the seventeen years I taught at our Nebraska field station; both students went to hospitals and recovered fully. Yet, both bites could have been avoided by the students taking the simplest of precautions; but teenagers in groups, especially teenage males, are even less predictable and often operate less logically than do rattlesnakes.

The Cassin's sparrow (Figure 40) is one of those essentially southwestern grassland species that periodically and unpredictably invades the central plains states, nesting with some regularity as far north as southern and western Nebraska. But its primary habitat is the mesquite grassland region of western Texas. In dense mesquite cover, the birds are more likely to occur in edge areas, using the dense cover as refuge but foraging in more open grasslands. It is a large, rather pale sparrow, somewhat like the vesper sparrow in its attraction to dry sites; and also like the vesper sparrow its outer tail feathers are mostly white, providing a useful fieldmark in flight.

Cassin's sparrows are persistent singers, vocalizing from February

Figure 40. Cassin's (left) and Brewer's (right) sparrows, adult males

until September in Texas. Their song is a liquid trill that descends in pitch and is followed by two short notes. Like other arid grassland birds, the territorial song is often uttered during a song-flight display. Then the bird ascends to about twenty feet, begins singing near the apex, then sets its wings and parachutes back to earth or to a favorite perching post, singing throughout. The usual song consists of six phrases and lasts more than two seconds. It typically begins with one or two soft introductory whistles, followed by a descending trill and terminated by several rapid notes of varied frequencies.

The very prolonged singing period of males would suggest that persistent renesting is likely, but very little information is available. In one Texas study ten nests were found, of which nine were placed in prickly pear cactus. All the nests in cactus were placed close to the ground, and all were on the south or southeast side of a leaf pad, providing some protection from the heat of the afternoon sun. Several of the nests also contained brown-headed cowbird eggs, but the available sample sizes are too small to indicate the seriousness of this threat. During the period 1966–2000, the Cassin's sparrow declined at a national average rate of 2.3 percent annually on breeding bird surveys, and in the Central States Region of the USFWS it declined at an average rate of 2.4 percent during the same period.

The Brewer's sparrow (Figure 40) is a similarly small, inconspicuous sparrow of the sage-dominated shrubsteppe of the American west that might easily be overlooked were it not for the male's wonderful song. It is even less visually memorable than the generally drab Cassin's sparrow. Like the Cassin's sparrow's apparent attachment to mesquite, the Brewer's sparrow has an equally firm association with sagebrush, especially big sagebrush. The U.S. ranges of the sparrow and big sagebrush are nearly identical, and I have personally never seen the sparrow except in sage-dominated areas. This does not mean that the Brewer's sparrow eats sage, but the plant evidently provides the perfect physical cover for establishing breeding territories and nesting.

The territorial song of the Brewer's sparrow is surprisingly complex for such a small and inconspicuous bird. Its typical territorial song is the "long song," which is a prolonged series of buzzes, trills, and bubbling phrases on varied pitches that may last for ten seconds or more and may be continued almost without interruption for several minutes. Like wrens, the birds seem to make up for their small size and drab patterning by their vocal efforts. In addition to their spectacular long songs, the birds also utter "short songs" that are typically two-parted trills, the second half being at a lower pitch and often longer than the first. Most songs are uttered from sagebrush, but if taller shrubs such as juniper are present, these may be selected preferentially.

Even though big sagebrush once covered more than a hundred thousand square miles of western North America, the Brewer's sparrow population is seriously declining. The causes of this decline are uncertain, but widespread alteration and destruction of the sagebrush ecosystem is a likely contributing factor. During the period 1966–2000, the Brewer's sparrow declined at a national average rate of 3.1 percent annually on breeding bird surveys, and in the Central States Region of the USFWS it declined at an average rate of 4.6 percent during the same period.

Mice are of course common residents of scrub-dominated habitats, where they provide an important food base for hawks, owls, snakes, foxes, coyotes, and other grassland predators. The hispid pocket mouse (Figure 41) is one of several species of pocket mice that are common in many parts of the Great Plains, especially in sandy substrates where burrowing is easy. They are named pocket mice because of their fur-lined cheek pouches, used to carry dried seeds to the burrows; without the fur lining the seeds might become wet and prone to mold during storage. The hispid pocket mouse is additionally named

Figure 41. Hispid pocket mouse, adult

for the rather bristly appearance of its longer black guard hairs that contrast with an otherwise more generally yellowish pelage.

Hispid pocket mice are the largest of the Great Plains pocket mice, ranging up to nearly two ounces as adults, or as much as four times heavier than the smallest species. Over most of the year, the animals live exclusively on seeds, but in spring insects may constitute up to about 15 percent of the total diet. The seeds gathered are selectively chosen, and over the long winter months the animals rarely if ever venture out. Then their burrows are kept plugged with earth most of the time, keeping out at least some potential predators. They may also become dormant and estivate in hot, dry weather, although they are not believed to require freestanding water at any time. For most of the year, these animals live solitary lives, becoming social only for breeding.

Pocket mice and woodrats are the primary foods of diamondback rattlesnakes in Texas, supplemented mainly by kangaroo rats, harvest mice, pocket gophers, rock squirrels, voles, and rabbits. The diamondback rattlesnake is the largest of the Great Plains rattlesnakes, reaching maximum lengths of nearly ninety inches as compared with about seventy inches in the timber rattlesnake, forty-five inches in the prairie rattlesnake, and thirty-five inches in the massasauga. In central and eastern Oklahoma, it occurs in some of the same locations as the timber rattlesnake, but it extends farther west and is more likely to occur in rocky or mountainous habitats, such as the Wichita Mountains and Gypsum Hills. Instead of the black tail of the timber rattler or the banded gray and brown tail of the prairie rattlesnake, the diamondback's tail is pale gray to whitish with black rings.

Studies in southwestern Oklahoma have shown that this snake is most active during the cooler parts of the day, migrating during spring from winter dens to summer territories up to a mile distant and then returning to the same den in autumn. Males tend to move farther than females, averaging about 100 meters per day in spring as compared with about 60 meters per day by females. Males are sexually active throughout the summer, and females produce average broods of about nine young. There may be a very short period of parental attendance by the female after birth. The young are venomous from birth and soon begin to feed on their own. Small mammals such as woodrats and pocket mice are common prey, but birds and amphibians are sometimes also taken. Bites by large rattlesnakes of any species are serious and are sometimes fatal. Some diamondback rattlesnakes have reached ages of ten to fifteen years in the wild.

CHAPTER 7

Shaded Shorelines and Tall Trees

The Riverine and Upland Hardwood Forests

Dorothy may well have believed, all the way down to her ruby slippers, that there is no place like home on the dry plains of central Kansas, but I suspect she never saw a prairie river. I am thinking of such historic plains rivers as the Platte, the Republican, the Kansas, and the Cimarron, but most especially the Platte, the archetype of all prairie rivers. It was the Platte that the fur traders, the Mormons, the emigrants headed for Oregon, and the railroad builders who were determined to build a northern route to California all chose to follow. No other Great Plains river so neatly bisected the otherwise relatively waterless central plains, taking nineteenth-century travelers from such "civilized" starting places as Plattsmouth and Omaha almost all the way to Wyoming's North Pass, the lowest point in the continental divide north of New Mexico. Following the Platte west assured emigrants of a nearby source of green grass and water for their cattle and oxen, but the Platte's water was also a source of cholera during that period, and this terrible disease ravaged some immigrant trains.

When I was a graduate student in Ithaca, New York, and had not yet even set foot in Nebraska, I often listened with envy to the stories of other graduate students who had done fieldwork in the Platte Valley. Accounts of shady cottonwoods lining a lazy prairie watercourse, with brilliantly plumaged orioles, grosbeaks, buntings, and a host of other eastern forest birds meeting their western counterparts in the central Platte Valley along with bobolinks, song sparrows, and other wet meadow and edge species, seemed like an ornithological combi-nation that I must one day see. I hadn't even heard yet of the sandhill cranes that poured endlessly into the Platte Valley each spring, far more abundantly and vastly more beautifully than the flying monkeys in the Wizard of Oz.

Indeed, from my distant viewpoint in New York, Nebraska seemed to me to be a kind of fictional Oz itself, and when I finally arrived its natural beauties surpassed even my wildest imagination. Unlike the muddy Red River of my youth, the relatively clear Platte had a sandy bottom and was shallow enough to wade entirely across. Its century-old cottonwoods, elms, hackberries, and occasional bur oaks seemed to me to carry the secrets of uncounted thousands of travelers who must have rested in the shade of these very trees, or at least the shade of their immediate predecessors. I knew too that the Pawnees who lived along the Platte's shoreline considered many locations along it to be sacred, being home to special animal spirits. I was quite content to revel in the sights and sounds of the actual animals.

The riverine forests of the Great Plains might be thought of as nar-row, ameba-like arms extending out from upland forests widely dis-tributed on both sides of the plains. In the southeast, botanically diverse deciduous forests dominated mostly by oaks and hickories from geologically ancient regions, such as the Ozark Plateau and Oua-chita Mountains, snake westward along the Arkansas and Canadian Rivers and northwestward along the Missouri, Kansas, Republican, and Platte Rivers. The cold-adapted hardwood maple and basswood forests of central Minnesota likewise follow the Minnesota and Red Rivers northwestward into the eastern Dakotas. From the west, mon-tane coniferous forests dominated by ponderosa pines stretch out locally into the high plains from the Rocky Mountain piedmont, find-ing eastern expansion corridors along the upper reaches of the Mis-souri, Niobrara, Platte, Arkansas, Cimarron, and Canadian Rivers. And in western South Dakota, the Black Hills rise majestically above the grassy plains in solitary splendor, producing an island of pon-derosa pine and other conifers that in most ways biologically resem-ble the pine-dominated forests of the Bighorn and Laramie Mountains a few hundred miles to the west. In only a few places do the eastern hardwood forests extend westward far enough to meet their western counterparts to any great degree, but in the Niobrara Valley of Ne-braska this does indeed occur with fascinating ecological results, such as tree hybridization, bird hybridization, and strangely juxtaposed species involving such diverse woods-adapted bird groups as towhees, buntings, orioles, and woodpeckers.

Of all the Great Plains woodpeckers, the northern flicker is the one that seems best adapted to the Great Plains. It is as much a forest-edge species as a woods-dweller, for one of its favorite foods is ants, which it often obtains by flying out into grasslands near woods and probing for them in the ground. As a result, it is the most common breeding woodpecker in Nebraska and Kansas, slightly edging out the red-headed woodpecker in both cases. The same is likely true in both Dakotas, where at least possible if not proven breeding by flickers has been found in all their counties.

Because of its important numerical position in the Great Plains, the nesting holes that northern flickers drill in trees become extremely important as breeding sites for all those species of cavity-nesting birds that must depend on woodpeckers for providing their own lodgings. Species such as the eastern bluebird, tree swallow, chickadees, titmice, nuthatches, wrens, and others are all to varying degrees nest-site commensals of woodpeckers. Northern flickers, like the red-headed woodpecker, have suffered major population declines in recent decades, both species declining at annual rates of 2 to 2.5 percent annually. Their loss to a breeding avifauna is especially sad because of their role in providing nesting sites for cavity-dependent birds.

Beyond its special ecological role, the northern flicker is perhaps the best poster bird for representing the major role that the Great Plains have played as an evolutionary "suture zone" between eastern and western biotas. The western form of the flicker is the red-shafted, and the eastern version is the yellow-shafted, the differences between them being largely limited to the presence of yellow versus red carotenoid pigments in their tail and wing feathers. This convenient genetic marker provides a simple basis for judging the likely ancestry of any flicker specimen, and the intermediate color types indicate a likely hybrid origin. By detailed analyses of hundreds of flicker specimens, a "hybrid zone" stretching from western Canada to Texas can be traced, suggesting the approximate meeting points of eastern and western flicker gene pools. Similar hybrid zones have been established for several other pairs of forest-edge or woodland-adapted birds. In many instances, these zones are quite similar geographically, typically crossing the central and western parts of the region encompassed by this book and biologically showing us where east truly does meet west. Examples include the Baltimore and Bullock's oriole, the black-headed and rose-breasted grosbeak, the spotted and eastern towhees, and the indigo and lazuli buntings.

Indigo buntings are intensely blue in direct sunlight. In such a situation one can only imagine that they would attract every predator within sight of them as they sit on the tips of a tree branch in full sunshine, singing exuberantly. The brilliant blues of these birds are as much a result of light-scattering of the shorter blue rays of light (the same basis for our perception of blue skies) as they are of true iridescence, which depends on the refraction of light waves by the internal feather structure, as is produced by a prism. In either case, a relatively directed source of light is needed to produce the full visual effect. After showing my ornithology class male indigo buntings under perfect field viewing conditions and letting them think they had an easy bird to remember for future quizzes, I could hardly wait for an overcast day to test them. Then, I would point out a dingy grayish-appearing and sparrow-sized bird on the top of a distant tree and say, "OK, that's your next quiz bird." Unless they happened to remember its song, they almost never came up with the proper identification.

Indigo and lazuli buntings are among the songbirds in which hybridization is occurring where the two close relatives meet in the central plains, after having been geographically and ecologically separated for at least millennia, if not longer. The lazuli bunting is the western relative; instead of being almost entirely sky blue, breeding males are white-bellied and rufous-breasted. The lazuli bunting occurs from the central plains west to the Pacific coast and is adapted to somewhat more arid climates and more scrubby habitats than is the indigo bunting. Our field station in western Nebraska is located near the middle of the hybrid zone, and the incidence of bunting hybrids during some summers was so great that variously intermediate birds were more abundant than the usually common indigo or relatively rare lazuli bunting plumage types. This additional level of confusion didn't improve matters in terms of field quiz efficiency.

If the indigo bunting male can appear sky blue in bright sunlight, then so too can the eastern bluebird male. One early American nature writer described the male bluebird as carrying the sky on its shoulders. Historically, eastern bluebirds were one of the most common breeding birds along the western edges of the hardwood forests that gradually graded into tallgrass prairies. However, two major factors led to their near demise. By the 1930s, European starlings, which had been introduced from Europe into New England in the late 1800s, had spread west far enough to reach the plains states. They were first reported in Nebraska in 1930, and similarly reached Oklahoma in the winter of 1929–1930. By the end of the 1930s, they had crossed all of Nebraska and continued their march west to the Pacific coast. Starlings proved to be strong competitors with eastern bluebirds for available nesting cavities and probably initiated the bluebird's long decline. The appearance of "hard" organic pesticides such as DDT in the mid-1940s only hastened their decline, and by the time I reached Nebraska in the early 1960s bluebirds were already rare.

For many years during the 1970s and 1980s I never saw an eastern bluebird in the state and assumed they were gone forever. Yet, by the late 1980s a nationwide effort was under way to reestablish bluebirds by erecting and closely monitoring nesting boxes for them, and the tide slowly turned. I have now helped in this effort for about six years and can count on seeing bluebirds, their eggs, or their young almost every time I go out to check their nesting boxes.

Eastern bluebirds, like indigo buntings, are forest-edge birds, doing most of their foraging over grasslands and meadows, so bluebird nesting boxes should be placed near meadows or fields that offer foraging opportunities. As such, one might expect them to meet and perhaps hybridize with mountain bluebirds, which geographically replace them along the Rocky Mountain piedmont. Such hybridization is probably quite rare, but hybrids have been observed in Kansas, Nebraska, and southern Canada.

An added benefit to erecting nesting boxes for bluebirds has been the addition of tree swallows to the list of common nesting birds in the region, a species previously rather rare as a nester south of the Dakotas. Tree swallows will readily accept boxes built for bluebirds, and by placing nesting boxes in pairs rather than singly, both species can be induced to nest within a few yards of one another. It is hard to judge which species can give one more pleasure when working a "bluebird trail." The wonderful blues on the upperparts of a bluebird are fully matched by the iridescent blues and violets of a tree swallow's back.

Another hole-nesting species that typically relies on woodpecker holes for nest sites but will readily adopt nest boxes is the eastern screech-owl (Figure 42). Probably most residents of suburbs and cities

Figure 42. Eastern screech-owl, social preening by paired adults

of the plains states would scoff if one suggested to them that they might well have a pair of owls nesting in their back yard. Yet, such is often the case, for screech-owls are small, inconspicuous, and relatively silent birds. They do not "screech" except when under extreme duress; indeed only the barn-owl among Great Plains owls may be properly described as a real screecher. The usual call of the eastern screech-owl is a soft, quivering, and ghostly wail that might be easily overlooked among the many sounds of a city. Even if one hears and recognizes the call, finding the bird that made it is nearly impossible. The typical plumage color of Great Plains screech-owls is a mottled dark gray and buff that almost perfectly matches tree bark. When alarmed, a screech-owl will assume a slim, erect posture, lean against a vertical trunk or branch, and nearly close its bright yellow eyes, thus hiding it from view. It thereby merges with its background in a near-magical way, remaining immobile until the danger has passed.

In more eastern parts of the screech-owl's range, a significant proportion of the species exhibits a richer, rufous-toned plumage, which seems to match the more humid surroundings better than a gray one would. In the eastern parts of the Great Plains states, about 5 to 10 percent of the screech-owls are of this rufous morph, and some individuals are of an intermediate brownish-gray color. Other than in their plumage traits the birds are identical, and since it is possible that the birds are essentially color-blind, it may be that their individual plumage tone is of no social significance to them. West of the plains, in the Rocky Mountain piedmont, western screech-owls replace the easterns. They are nearly identical in plumage to the gray morph of the eastern, but do have recognizably different vocalizations. This seems to be yet another biological example of east meeting west in the Great Plains.

Screech-owls are omnivores, taking a diversity of insects and small vertebrates such as nestling birds when they are most easily available during summer, but gradually shifting over entirely to vertebrates during fall and winter. They are sedentary birds, probably rarely leaving their home ranges once they have gained independence from their parents. They largely depend on the holes made by woodpeckers for nest sites, but natural cavities are used as well. At least in Nebraska they are perhaps the most abundant of all the owls, but great horned owls are much larger and more conspicuous. Great horned owls are probably also the worst of the screech-owl's enemies. There is no honor among owls in terms of what represents fair game; larger owls eat smaller owls, and the great horned is the largest and most dangerous of all.

Compared with the screech-owl, which weighs in at about seven or eight ounces, the great horned owl is a giant, averaging about three pounds in males and four pounds in females. Because of its great strength and power, it is not nearly so shy about revealing its presence to humans and other animals, for few other predators would willingly take on one of these impressive birds. A great horned owl would not hesitate to attack a large rattlesnake, a pet cat, or even a small dog, should it get the chance. Each of its toes and stiletto-sharp talons can produce a viselike grip representing nearly thirty pounds of pressure, a grip from which few animals can escape alive.

Great horned owls are common everywhere in the Great Plains. They easily represent the largest number of owl breeding records for Nebraska and Kansas, and there are "possible" to "confirmed" breeding records for nearly every county in the Dakotas. I have estimated that great horned owls alone represent about a third of all the records for the roughly dozen owl species that I was able to accumulate and summarize for the state of Nebraska. They are the strigid counterpart of the red-tailed hawk, which has a corresponding premier position of abundance among all the hawks of the state. Both are large and powerful predators that are able to exploit wide, broadly overlapping prey bases. They avoid direct competition by differences in their activity patterns; the red-tails rule the day, while great horned owls operate between dusk and dawn. Only during the darkest hours, when limitations of the great horned owl's eyesight and hearing make hunting unprofitable, is it fairly safe for barn owls and barred owls to be active.

A few years ago a great horned owl took over a Lincoln nest of a red-tailed hawk that belonged to a resident pair that had used it for several years. Since owls begin nesting very early, sometimes as early as late January during a mild Nebraska winter, this pair of owls was well into its breeding before the red-tails probably knew they had lost their home. The owls either fledged or lost their young by late March, when they abandoned the nest and let the red-tails take possession. One unhatched owl egg remained in the nest and was incorporated into the red-tailed hawk's developing clutch. Remarkably, this egg hatched at about the same time as did the baby red-tails, and the owlet was fed and reared by the hawks as if it were one of their own. It evidently fledged with the hawks, but almost certainly had become imprinted on them, as baby owls are very prone to accept foster parents as potential mating partners when they mature. It would be nice to know whatever became of this bird, but it is unlikely that any red-tail would ever allow an adult great horned owl to approach close enough to court it!

The red-tailed hawk (Figure 43) is the so-called "chicken hawk" of my childhood, a generic vernacular name for the large, broad-winged, and wide-tailed hawks that are the signature daytime aerial predators of the Great Plains. Few chickens are ever eaten by red-tailed hawks; they primarily hunt small mammals. They are the larger hawks most likely to be seen perching on telephone poles, looking for mice lurking in the mowed ditches, but also willing to take a chance on almost

Figure 43. Red-tailed hawk, adult

anything that looks vulnerable. They are a good deal more wary than Swainson's hawks, which are also common pole-sitters, and can be rather easily separated from them at takeoff. If the rusty-red upper tail color of adults is not apparent, then the mostly white underwing color, including the longer wing feathers, should help distinguish red-tails. Swainson's hawks have a grayish upper tail and a rather uniformly dark underside color to the long primary and secondary feathers. Rarely, very dark, even almost black plumage variants of both species occur, making field identification much more difficult and requiring less obvious comparisons of relative wing shape differences.

Red-tailed hawks average about 2 to 2.7 pounds as adults, the females being somewhat larger than males. They are also larger than adult Swainson's hawks but smaller than ferruginous hawks, their most common ecological associates on the Great Plains. Red-tails are the most tree-dependent of these three species, gradually and sequentially giving way to Swainson's and ferruginous hawks as one moves west into the mixed-grass and short-grass plains. All are deadly predators on rodents, and a pair of red-tails nesting in a farm lot is almost as valuable in controlling mice and rats as a pair of barn owls.

Like all the avian predators, red-tails are strongly monogamous, and the same pair will regularly use the same nest site year after year. In our region the birds are also relatively sedentary, although fall and winter bring into the Great Plains many red-tails from much farther north in Canada and even Alaska. It is these birds that most often show unusual and confusing plumage variations, such as blackish or melanistic individuals. As one proceeds south from North Dakota, progressively more red-tails overwinter in the Great Plains, with the numbers staying on the Oklahoma plains sometimes truly amazing. By that season the Swainson's hawks have migrated to South America, making field identification of red-tails much easier.

Cuckoos are a bit like screech-owls, in that many Great Plains residents have heard them frequently, but few would claim to have seen one. Most people expect a cuckoo to sound like a European cuckoo clock, not the rather wooden-sounding and repetitive notes that emanate from the edges of woods on cloudy summer days or toward evening. Such sounds are thought to be produced by "rain-crows," rather mythical Great Plains creatures perhaps akin to thunderbirds. Most listeners would be surprised to learn that the rain-crow is nothing more (or less) than a yellow-billed cuckoo.

There are actually two species of cuckoos in the Great Plains, but the yellow-billed is the more common one except in South Dakota, where the black-billed predominates, and in North Dakota, where only the black-billed species occurs. Both are similar in appearance, differing in major part by their bill color. Both also feed preferentially on caterpillars, especially the very hairy ones such as tent caterpillars that are avoided by nearly all other birds. They skulk about in heavy vegetation, rarely showing themselves for long, and are far more often heard than seen. They also are one of the last birds to arrive in spring, usually not before late May, by which time a goodly supply of large caterpillars is likely to be available. It is not unusual for the remains of several hundred hairy caterpillars to be present in the stomach of a single cuckoo, the lining of which is sometimes made "furry" by the countless hairs that become impaled and incorporated into its surfaces.

Nests of both cuckoos are rather flimsy affairs, and both their nests and bluish-green eggs are almost indistinguishable from one another. Perhaps this is why yellow-billed cuckoos sometimes lay eggs in the nests of black-billed and vice versa. However, nest parasitism of other species by either of our native cuckoos is relatively rare. The three or four eggs are incubated as soon as they are laid, so that the young may hatch at intervals of several days after about nine to eleven days of incubation. The nearly naked, black-skinned, and rather primitive-looking chicks may begin to clamber about on branches when only a week old. Like trogons, the developing body feathers remain in their sheaths until nearly fully grown, producing a strange prickly or "porcupine" nestling stage until they are close to fledging. They are unable to fly until they are at least three weeks old, a very long fledging period but perhaps understandable in light of the very short incubation period.

The yellow-billed cuckoo as well as its relative the black-billed cuckoo are two of the many Nebraska breeding species in serious population decline, both dropping at rates of about 1.8 percent annually between 1966 and 2000. Most of the other Nebraska species in even sharper declines are grassland breeders, such as Henslow's, grasshopper, and lark sparrows, Sprague's pipit, horned lark, and eastern meadowlark. The cuckoos are birds of woody riparian edges and hardwoods, so their recent declines are harder to understand.

Cuckoos are also unusual birds in that they typically exude a strange, somewhat repugnant odor when handled. However, compared with skunks, cuckoos are pleasant company. The common striped skunk (Figure 44) is fairly familiar to anyone who spends much time in the field or even drives down country roads, where the stench of a road-killed skunk is sometimes hard to avoid. Probably there would be far fewer skunks killed by autos if these animals hadn't evolved their own usually effective defense mechanism. This remarkably uniform and powerful strategy has allowed all skunks to lose whatever fear they might otherwise have had relative to large moving objects. Their distinctive black-and-white pelage pattern should serve as an effective reminder to any foxes, coyotes, and other mammals that have had unfortunate prior encounters with a skunk, but it doesn't seem to work against great horned owls, very hungry bobcats, or Volkswagens.

Adult skunks can store about three teaspoons of vile-smelling musk in their anal glands, enough for five or six discharges. Because of their limited supply of such valuable ammunition, skunks are judicious in their use of it and would rather resort to a threat display of raising and waving their tail, foot-stamping, or hissing to warn intruders that there may be worse things to come. Their effective maximum range is up to almost twenty feet, but they are fairly accurate only at distances of less than half that. I once had a small terrier that, during a hike through the Minnesota woods, flushed a skunk from a hollow log. In spite of my frantic screams to my normally obedient dog, he acted like any good terrier, ignored me, and attacked. The attack didn't last long, and for several weeks thereafter the dog slept outside. At least he was much more obedient after that. Like the scent of spotted skunks, this odor is hard to eliminate from clothing, but washing in gasoline, ammonia, or dilute laundry bleach such as Clorox may be helpful. Application of a mixture of 3 percent hydrogen peroxide, baking soda, and a little liquid detergent helps remove it from a pet's fur but may turn your pet into a blonde. I gave my dog the traditional treatment of canned tomato juice, which at least changed its appearance if not its smell. Tomato juice simply masks one odor with another.

Unlike some larger mammals such as woodchucks and marmots, skunks do not truly hibernate, but during very cold weather (about 15° F for males, below freezing for females) they may become somewhat dormant. At that time, their body temperature remains high, but the animals may become inactive for as long as a month or so in adults and several months in immatures. Like true hibernators, the animals

Figure 44. Striped skunk, adult warning display

lose a good deal of weight at such times, up to as much as 40 percent in the case of females.

Skunks are essentially omnivorous, eating almost anything they stumble across, including insects, shrews, rats, and mice. Larger mammals may be eaten as carrion, but small ones such as mice are typically caught alive. There are two very similar species of mice with white feet and deer-brown upperparts that are widely distributed in the Great Plains woods and nearby grasslands, the deer mouse and the white-footed mouse (Figure 45). Their large ears and bulging eyes provide evidence of their nocturnality, and in their diets they are about as generalized as any small mammal. As a result they are usually abundant, the populations often ranging up to about sixteen animals per

Figure 45. White-footed mouse, dozing (above), and nose-to-nose encounter behavior by male and female (below; after photo by J. F. Eisenberg)

acre. Both are the most common of all native mice in North America and are very similar to one another.

Their genus (*Peromyscus*) includes at least fifteen species north of Mexico, with the deer mouse and white-footed mouse having the broadest ranges. The white-footed mouse is more likely to occur in wooded areas, especially hardwood and mixed forests of the eastern states, but follows the riverine gallery forests westward all the way to the Rocky Mountains. The deer mouse is even more widespread and more diverse ecologically, with nearly sixty described subspecies. The open country races of deer mice, widely distributed and abundant in the Great Plains, tend to be somewhat smaller than white-footed mice and have shorter but more distinctly bicolored tails. Both species are largely nocturnal, communicating by scent signals (pheromones) and vocalizations that are partly ultrasonic, above the range of human hearing. However, they do utter squeaking, chittering, and shrill buzzing sounds that may be heard by humans from as far away as fifty feet. They will also stamp their feet rapidly when excited. During daytime the animals may doze, and in the coldest parts of their range they may become torpid when food is scarce, adopting a state of semi-hibernation. During torpor their rate of breathing may drop to about sixty per minute and their body temperature to about 60° F.

White-footed mice typically have home ranges of about an acre or so, but rarely these may be as large as ten acres. Males are more mobile than females, and some have been known to return to their home area when released from as far away as two miles. *Peromyscus* mice are preyed upon by a great array of predators, and it has been estimated that less than 20 percent of those born survive to sexual maturity, which occurs in about two months. Animals born early in the spring may thus breed in late spring, those born later may breed in the fall, and those born in the fall will not breed until the following spring. During the breeding season, the animals live solitarily or in loose pairs, but should a mate die, a new one is quickly acquired. During winter a dozen or more mice may share the same nest and will often huddle together to remain warm. During fall they stockpile foods, such as seeds and nuts, to help them get through the difficult winter period. Such stockpiles may be in old bird nests, in trees, or underground.

There are also two very similar species of jumping mice (Figure 46) in the Great Plains, the meadow and the western. Like *Peromyscus*, one (the western) is slightly larger and its tail is not so strongly bicolored.

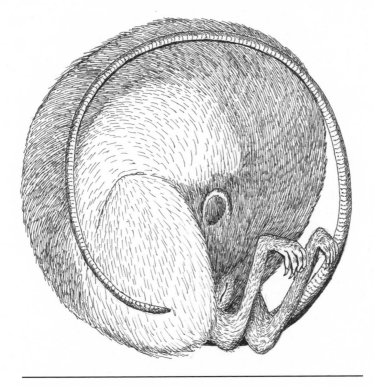

Figure 46. Western jumping mouse, adult hibernating

However, both species have very long tails and distinctly enlarged hind legs, which accounts for their jumping abilities. Both species are fond of rather moist habitats, such as low meadows and riparian woodlands with well-developed understories of grasses and forbs. They usually move about by making short hops in a zigzag direction, generally covering only a few inches under normal conditions. However, when frightened, the distance covered by single leaps may be several feet and on rare occasions, up to ten or twelve feet. In very good habitat their densities may reach nearly twenty animals per acre, with individual home ranges of up to about four acres. The animals are solitary, with variable and overlapping home ranges. One of their favorite food types is fungi, which they often find by digging. They also eat insects, green vegetation, fruits, and seeds. Like harvest mice, each of their upper incisors has a deep groove on its front surface.

Besides their jumping and digging, they are also proficient climbers and swimmers. They can even dive underwater to some depth and remain there for as long as a minute. Their unusually long tail probably not only serves as a useful counterbalance with jumping but also provides a sound signal by tail-drumming. However, it is not used for propulsion during swimming, in the manner of a muskrat.

Jumping mice are nocturnal and, unlike *Peromyscus* mice, enter a prolonged state of hibernation. In the northern plains, hibernation lasts from October through April in the case of the meadow jumping mouse, and from September to May or even early June in the case of the western species. During hibernation there seems to be a high mortality rate, perhaps especially among those animals that enter it in less than optimum condition. Winter nests are located well below the soil surface, sometimes more than two feet below, and consist of ball-like accumulations of dead grass and leaves. Within this nest the hibernating animal also forms a ball-like shape, with its long tail curled around its body. Assuming a spherelike shape results in the minimum surface area relative to volume and keeps the rate of body heat loss to a minimum. Sometimes two animals will occupy the same nest. During hibernation their body temperature may range between 35° and 40° F, and at the start of hibernation as much as two-thirds of the body weight may be represented by fat. Most of this is lost early in hibernation, when the body temperature is dropping most rapidly. After the fat stores have been reduced to 15 percent, the animals must either wake up and perhaps face cold weather, or continue to lose weight and perish in their sleep. A surrounding temperature of 47–48° F will stimulate spring arousal. Occasionally a hibernating mouse will rouse and may even move about a bit before falling back into dormancy.

Of all the nearly two dozen bat species that occur in the Great Plains, the eastern red bat (Figure 47) has the broadest overall range, from the prairie provinces of Canada south to Mexico, and its western counterpart extends to South America. It is a medium-sized bat with long, relatively pointed wings, perhaps indicative of its mobility and migratory tendencies, at least at the northern edge of its range. This so-called "red" bat actually has a wide range in pelage colors, the males varying from a distinctive bright orange or brick-red to rather yellowish, while the females are a duller chestnut buff. Both sexes have a contrasting whitish-yellow patch on each shoulder that sometimes is continuous across the chest. Most adults weigh between a

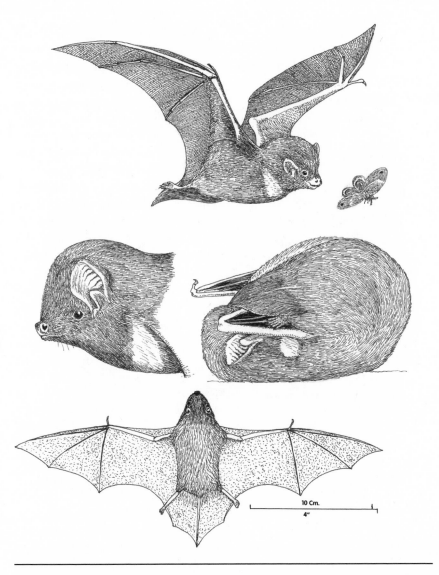

Figure 47. Red bat, in flight (above), head detail and hibernating posture (after a photo by R. W. Barbour), and wing-profile outline of adult male (bottom, from specimen).

quarter ounce and a half ounce, the females averaging slightly larger than the males.

The red bat is mostly migratory in the Great Plains, but some apparently hibernate in certain areas. During hibernation the animal curls up, the well-furred tail membrane pulled up like a blanket to cover its nose. This is typically done while hanging by the hind legs, but when placed in a refrigerator, a dormant red bat resembles a hibernating mouse. Prior to fall migration, the animals put on considerable fat, just as migratory birds do, and likewise migrate mainly at night. Sometimes they cross large bodies of water, and migrant flocks have not only been seen as far as 100 miles out in the Atlantic Ocean, but individuals have also reported sightings from as far away as Bermuda, nearly 600 miles from the coast.

Like other insectivorous bats, red bats use ultrasonic vocal signals to navigate in the dark and to locate moths and other insects, their primary prey. During daylight hours they typically roost solitarily in trees or tall shrubs, where they resemble dead leaves. They often roost in wild plum bushes, where they sometimes are mistaken for ripe plums. Likewise females do not form nursery colonies, but instead tend to their youngsters alone. Up to four young (usually two) may be born, and these are carried about by the mother for about a week after birth. Thereafter the young are left at the roost when the mother leaves for her nightly foraging, which is often done under streetlights. This is one of the very few North American bats known to give birth to more than one young at a time and the only one with brick-red fur.

CHAPTER 8

The Western Rim of the Plains

Coniferous Forests and Woodlands

From the conifer-covered uplands of the Rocky Mountains, tendril-like extensions of western forests historically have penetrated eastwardly into the Great Plains, especially along the rivers flowing down the slopes of the Rocky Mountain piedmont. Many of these tree and shrub species of the cordilleran forests were probably eliminated when they encountered the stark aridity of the high plains, but others have survived and maintained precarious existences. This has been the case especially along the shady north slopes of steep-sided river valleys, or at elevations just high enough to intercept what little moisture was left in the prevailing winds that were largely squeezed dry as they crossed the high peaks of the continental divide. Here, western forest trees such as ponderosa pine and Rocky Mountain red cedar could compete directly with prairie species, especially where the trees were protected from periodic fires. Here too arid-adapted shrubs such as junipers, mountain mahogany, and skunkbrush sumac sometimes provided a shady woody understory. In extremely dry sites, sometimes only hardy limber pines survived as scattered forest sentinels, but in cooler and higher locations the ponderosa pine stands became dense and tall and were supplemented by more cold-adapted tree species such as white spruce, paper birch, and quaking aspen.

As the western forests moved out into the plains, they were followed by many animals typical of the Rocky Mountain region. Not all of them made the journey, but most were helped by the periodic glaciations of the Pleistocene, when colder and wetter climates allowed montane forests and their associated animals to survive at lower altitudes, only later to become marooned on high peaks and steep slopes when the last of the glaciers finally retreated less than 20,000 years ago. This is probably how such northern forest mammals as northern flying squirrels and montane meadow species such as yellow-bellied marmots found their way to the Black Hills.

The Black Hills were never exposed to direct glaciation during the Pleistocene epoch, but the postglacial climate occurring then probably favored the development of a cool, locally dense coniferous forest. Both woolly mammoths and the even bigger Columbian mammoths wandered the Hills as recently as 26,000 years ago. There was also a variety of large predators, such as giant short-faced bears that were even larger than our present-day grizzly bear. Since then, the region's climate has become both warmer and drier, and many of the forest-adapted plants and animals now otherwise occurring only hundreds of miles to the west or north became isolated in the Black Hills, surrounded by high plains prairies and shrubsteppe vegetation. Similar isolations occurred in the Bighorn Mountains of north-central Wyoming and on several small, variably elevated areas of central Montana.

The situation just described concerns only relatively recent events; the whole story of isolated western coniferous forests and woodlands in the Great Plains began long before. Perhaps as early as about 60 million years ago there was a major uplifting of ancient layers of the earth's crust in the region of the northern Great Plains now represented by western South Dakota. As the granite, quartzite, and associated ancient materials dating from as early as pre-Cambrian times pushed upward, the sedimentary limestones that had been deposited immediately above them when the land was covered by Paleozoic seas were also uplifted, as were the higher and more recent layers of sediments. As these surface layers were increasingly exposed, their softer materials were progressively eroded by water, ice, and wind. Eventually, more than 6,000 feet of surface sediments were thus removed, gradually revealing the progressively older and harder core materials. Thus the rounded granite domes of peaks like Mount Rushmore and the sharper spires such as those called the Needles eventually came to dominate the central parts of the Black Hills.

Because of peripheral erosion, these ancient core materials are now surrounded in sequential and roughly circular fashion by progressively more recent geological layers, something like a mushroom whose convex top has been broken away to reveal its harder central stem. Water from rain and snowmelt also seeped down into the increasingly exposed limestone layers, slowly dissolving them and producing hundreds of small to large caves, including Wind Cave and Jewel Cave. Likewise, streams have cut new courses through the surface layers, most of them flowing eastward to join the Missouri River. Along the way, erosion produced by the Cheyenne River and its tributaries eventually formed the South Dakota Badlands.

Because of this ecological and geological isolation, populations of some plants and animals now occur in the Black Hills but almost nowhere else in the Great Plains. Among the more conspicuously isolated trees are small, local populations of limber and lodgepole pines as well as white spruce and several birches. Among birds, there is a unique Black Hills race of the dark-eyed junco, sometimes called the white-winged junco. There are likewise distinctive Black Hills races of the red-bellied snake, with isolated eastern relatives, and the wandering garter snake, with western affinities. Isolated Great Plains populations of the long-tailed vole and yellow-bellied marmot, both rodents typical of cool Rocky Mountain forests and high mountain meadows, similarly occur in the Black Hills. An isolated population of a generally boreal rodent, the southern red-backed vole, also occurs here. Local Black Hills races of the long-tailed weasel, thirteen-lined ground squirrel, least chipmunk, and red squirrel are likewise present.

Several species of bats inhabit the caves of the Black Hills, including the otherwise rather rare northern myotis and a local race of the widespread fringed myotis. The long-legged myotis is probably the most common bat, but in South Dakota the Townsend's big-eared bat, the western small-footed myotis, and the long-eared myotis are all largely limited to the Hills region. A local race of the bighorn sheep once occurred in the Black Hills and Badlands of both Dakotas, but was hunted to extinction early in the past century. A bighorn subspecies native to the Rocky Mountains was reintroduced into the Hills during the 1920s, but this population eventually died out. After later introductions, bighorns have survived locally in rather small numbers, mainly in Custer State Park. After the once countless herds of elk on the plains were eliminated, this species was also reintroduced into protected areas of the Black Hills, especially around Wind Cave and Custer State Park. Mountain lions and black bears have somehow survived their long periods of persecution, and at least the former still exists in small numbers in the Black Hills, their last major refuge in the entire Great Plains. The mountain lion population seems to be increasing, but the current status of the black bear is more questionable.

More than twenty species of birds breed in the coniferous forests of the Black Hills but almost nowhere else in the Great Plains, except perhaps in Nebraska's nearby Pine Ridge and in other local pine forests of western Nebraska and the Dakotas. Some are directly dependent on pine seeds as a source of food, such as the Clark's nutcracker and red crossbill. The species that prefer to nest in pine-dominated habitats but are not necessarily dependent on pines as a food source include the evening grosbeak, ruby-crowned and golden-crowned kinglets, red-breasted and pygmy nuthatches, Swainson's thrush, Townsend's solitaire, yellow-rumped and MacGillivray's warblers, plumbeous vireo, and the gray and pinyon jays. There are four woodpeckers that breed in the Black Hills but occur little if at all elsewhere in the Great Plains. They include the black-backed, three-toed, and Lewis's woodpeckers and the red-naped sapsucker. The first three are to varying degrees associated with recently burned areas of forest, where bark beetles are usually abundant. The sapsucker is more closely associated with aspens than with pines, as is the ruffed grouse, another Black Hills isolate. The northern goshawk and northern saw-whet owl are conifer-adapted predators having generally broad distributions in northern North America, but their breeding ranges extend south in the Great Plains to the Black Hills. The Cassin's finch and dusky flycatcher are also Black Hills isolates, the finch being a rare breeder in pines and the flycatcher associated with scrubby or open woods.

Many additional Rocky Mountain species, such as the mountain bluebird, western tanager, ovenbird, and pine siskin, are more common in the Black Hills coniferous forests than anywhere else in the Great Plains. Historically, the conifer-dependent blue grouse also occurred in the Black Hills, but was extirpated rather early. Topography-dependent endemic birds of the Black Hills and adjoining areas that need cliffs for nesting include the canyon wren and white-throated swift, whereas steep, scrubby hillsides are used by the Virginia's warbler and shady forest valleys by the cordilleran flycatcher. The American dipper breeds only along the swiftest mountain streams of the northern Black Hills.

Besides the ponderosa pine forests of the western parts of the Dakotas and Nebraska, a small intrusion of coniferous woodlands dominated by arid-adapted pines and junipers extends eastward from northeastern New Mexico into extreme western Oklahoma. This woodland type is best developed in a portion of Oklahoma's Cimarron County called the Black Mesa. Here surface layers of volcanic basalt and sandstone form flat mesas that have been eroded into deep canyons by the Cimarron River. Pinyon pines that produce relatively few but unusually large, nutlike seeds ("pine nuts") occur here, as do several junipers including Rocky Mountain red cedar. The nutritious pine seeds offer a primary food resource for some mammals and birds, and the juniper berries likewise represent important winter foods.

Many unique canyon- or woodland-adapted rodents occur in this small but biologically rich area, such as the rock squirrel, Colorado chipmunk, Mexican and white-throated woodrats, and rock, brush, and pinyon mice. Pinyon jays and western scrub-jays likewise harvest and cache pinyon seeds whenever they become seasonally available. Among other breeding birds, the oak titmouse, bushtit, canyon wren, and common raven are especially characteristic, the former two species being small chickadee-like birds that glean insects from tree bark and foliage, and the latter two being cliff- or canyon-dependent breeders. Within the geographic limits of this book, nearly all of these mammals and birds are confined to this small but biologically rich region.

Once occurring in uncountable numbers on the high plains and more eastern prairies, the elk (Figure 48) of the Great Plains are now largely restricted to large parks or sanctuaries. Within the Black Hills there are more than 1,000 in Custer State Park alone. There are also small and variably confined populations in North Dakota, Nebraska, Kansas, and Oklahoma. Except for the moose, the elk is North America's largest species of deer and was highly important to plains Native Americans, who knew it as the "wapiti." Adult males of the Rocky Mountain race average about 700 pounds and females about 500 pounds, a bigger sexual difference than occurs in the other plains-dwelling deer. Adult males also have by far the largest antlers of these deer, a fact related to the life-threatening fights between mature males during the rutting season.

Once widespread on the plains grasslands, elk were probably more common in eastern Nebraska than in the more arid west, since grasses

Figure 48. Elk, adult male

and other herbs were taller and thicker, and there may have been less competition from bison. As late as the 1860s, there were still "magnificent herds" to be seen along the Niobrara and Elkhorn Rivers, and "immense herds" occurred farther west. However, the last individuals in the state were killed in the early 1880s, about the time the last bison disappeared. Reintroductions in the early twentieth century, in places like Fort Niobrara National Wildlife Refuge and in the Pine Ridge, have brought the state's population back into the hundreds, if not the thousands. Elk in some nearby regions, including Wyoming, have been found to be infected with chronic wasting disease, a lethal brain malady that is closely related to mad cow disease and that is a serious threat to hoofed animals and perhaps also to humans.

Early fall in the Black Hills and other areas wherever elk occur is marked by the daily buglelike calls of males, especially during early mornings and late afternoons. By such calls and visual displays the males try to attract and gain control of as many females as possible and try to intimidate other males that might wander too close to their collection of potential breeders. A few years ago, a Nebraska wildlife photographer decided to try for a close-up shot of a threatening elk by playing the tape-recorded call of an elk bugle while parked in a large pasture where a small herd of elk was on exhibition. He got a much greater effect than he expected, namely several large antler holes in the side of his car's body, made by a charging elk that quickly responded to the recorded sounds.

In the autumn of 2001, I took a trip to the Black Hills to reinforce old memories. While driving through Custer State Park one evening, I came upon a group of six bighorn rams (Figure 49) sauntering down the side of the highway, pausing occasionally to nibble roadside grass. I had already seen nearly thirty females and juvenile males foraging a few miles away, but the sight of six magnificent rams practically within touching distance was almost unbelievable. Before I could stop and frantically replace my telephoto with a shorter lens, one of the rams suddenly stopped and struck a nearby male with a full, lunging horn-to-horn blow that nearly shook my car. It made me realize that these are not tame animals, and that such a sight would have been unheard of the first time I visited the Black Hills in the 1940s.

Like the elk, bighorn sheep once had a considerably broader range than they now occupy, but they have always been associated with steep, rocky terrain. They were originally actively hunted by the Mandans,

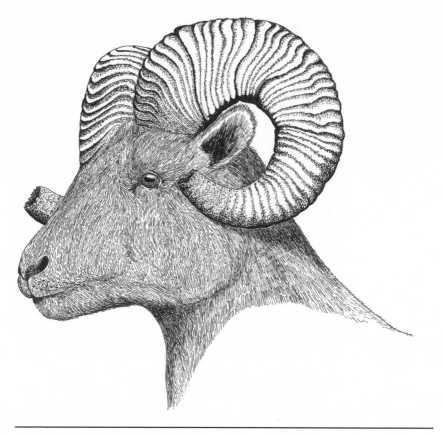

Figure 49. Bighorn sheep, adult male

Minnetarees, and perhaps other upper Missouri Valley tribes of the Dakotas, who made clothes from their skins and bowls or ladles from their horns. Among European settlers, they were not only regarded as a savory food source, but the massive horns of males also made them highly attractive trophies. As a result, the endemic Audubon's bighorn (named after J. J. Audubon) of the Dakota badlands and Black Hills was eradicated quite early. Explorers of the first half of the 1800s, including Lewis and Clark, Prince Maximilian, and J. J. Audubon, found the animals to be abundant along the upper Missouri River in what is now North Dakota. Theodore Roosevelt killed a ram along the Little Missouri River in the 1880s. It may have been one of the last surviving sheep in North Dakota; the last historic state record was of one killed in

1905. The Rocky Mountain race of the bighorn has since been introduced successfully into the North Dakota badlands.

The story in South Dakota is similar; bighorn sheep were eliminated from that state by the early 1900s, with the last confirmed record for 1905, on Magpie Creek. However, in 1923, releases began in Custer State Park, but this herd died out because of disease. Later successful releases began in 1965. A herd was started in Badlands National Park in 1963, about the same time that some confined animals in Harding County escaped and started a free-ranging herd in the Slim Buttes area. The Black Hills population numbered at least 150 animals in 2001, and there is a separate population in the park.

Also native to the Pine Ridge and Wildcat Hills of Nebraska, some bighorns historically occurred as far east as Lincoln County along the breaks of the North Platte River and east to the vicinity of Long Pine in the central Niobrara Valley. The last ones were extirpated from Nebraska by about 1890, but there have since been periodic incursions of a few animals from eastern Wyoming. Several reintroduction efforts were made in the Fort Robinson area of the Pine Ridge between 1981 and 1993, and in the Wildcat Hills during 2001. The state's population as of 2001 numbered about ninety animals.

Among the avian raptors of the Black Hills, the northern goshawk (Figure 50) is at the upper end of a sequence of stair-step–sized, swift-flying raptors, as also occurs among the North American falcons. In the case of the goshawk, the Cooper's hawk and sharp-shinned hawk are the progressively smaller versions, whereas in the falcon sequence the peregrine, prairie falcon, merlin, and American kestrel comprise the corresponding diminishing sequence. The goshawk and its smaller congeners all have short, rounded wings adapted for erratic maneuvering through trees and brushy vegetation, whereas the falcons have relatively pointed wings adapted for straight, open-country attacks. In either case, these birds bring sudden terror and sometimes also quick death to any smaller birds. In fact, larger birds are not secure, as goshawks will not hesitate to attack birds weighing more than themselves (about two to three pounds), even if they cannot carry them off. Domestic poultry weighing up to five pounds and woodchucks weighing seven to ten pounds are killed on occasion. There have even been accounts of goshawks attacking and trying to carry off wooden duck decoys.

George Sutton once related that, while watching a goshawk nest, he was repeatedly attacked by the female, the bird raking him with

Figure 50. Northern goshawk, adult

her powerful hind talons each time she swooped past. In spite of her size, she was so well camouflaged by the vegetation that on three occasions Sutton failed to see her before she struck him. Only by wearing a heavy cap and a piece of cloth wrapped securely around his neck did he avoid serious injury. I too was once similarly raked by a ferruginous hawk when I approached a confined but full-winged bird to photograph it. Luckily, it was a somewhat glancing blow to my forehead, but I might easily have been blinded by the sudden attack.

One of the favorite prey of goshawks is the ruffed grouse (Figure 51). In one analysis of their prey by George Sutton, 55 out of 251 goshawk stomachs contained the remains of ruffed grouse, or about the same number as contained cottontail rabbits. This study was done in Pennsylvania, and nothing is yet known of preferred goshawk prey in the Black Hills. However, goshawk prey types are generally similar everywhere, consisting mainly of grouse and rabbits or hares. Ruffed grouse are still fairly common in the Black Hills and also in North Dakota's Turtle Mountains, but were eliminated from some other historic Great Plains habitats such as the upper Missouri Valley. However, they have been reintroduced successfully into eastern Kansas.

The Black Hills and Turtle Mountains are quite different forests

Figure 51. Ruffed grouse, adult male drumming display

botanically, the former dominated by conifers and the latter by hardwoods, but both have one important element in common, namely aspens. Aspens and related tree types such as birches are critical elements of ruffed grouse habitat almost everywhere. Through the winter months their buds provide nutritious foods above the snowline, and in the spring their catkins offer females high-protein foods at the time of their maximum energy needs, just prior to nesting.

There are few nature activities more exciting then trying to see the spring drumming performance of ruffed grouse. Unlike the prairie grouse, male ruffed grouse are relatively solitary during territorial display. They typically select a rather large, somewhat rotted log lying almost horizontal on the forest floor, in an area of forest with fairly open understory vegetation. This site provides them with adequate visibility to detect approaching terrestrial predators and makes it harder for humans to witness the event. Digging their claws into the soft bark and rotted wood, the male stands crossways on the log, his tail spread and pressed down on the log to provide additional tripodlike support for his legs. Then, in a series of increasingly rapid wing-strokes, he brings his wings forward until they nearly touch, producing a soft, muffled sound that reflects sudden air-pressure changes rather than the wings striking one another directly. In spite of the seemingly weak sound that is generated, it carries for fifty yards or more under favorable conditions, and probably not only attracts females but also deters other males from approaching.

I have often tried to approach such "drumming" males by moving quietly toward them only during the brief drumming sequences and remaining absolutely still during the intervening intervals. With great care and patience it is often possible to get close enough to actually see the displaying male before eventually being discovered, which always brings a quick end to the episode. It is actually much easier to construct a blind for seeing these activities, but not nearly so challenging.

Just as the hawks lead the list of avian raptors that are active during the day, the owls progressively take over as night falls. To some degree all owls can hunt under reduced light conditions, but the acuity of their hearing and the light-gathering power of their eyes largely determine the degree to which owls can hunt during the darkest period of night. As suggested by their large facial disks, northern saw-whet owls (Figure 52) have remarkably good hearing. Their internal ears are also asymmetrically enlarged, apparently to improve binaural sound per-

Figure 52. Northern saw-whet owl, adult

ception, to such a degree that even the shape of their skulls has been modified. It is thought that such asymmetry enhances the owls' ability to localize sound in a three-dimensional environment. Although their name was based on the idea that their territorial and mating calls sound like the noise of a cross-cut saw being whetted, the calls are actually more evocative of the sound produced by dripping water.

147

Figure 53. Colorado chipmunk, adult

All of the North American owls are visually appealing, but few if any are more attractive than the saw-whet owl. Saw-whets have a look of perpetual surprise, their yellow eyes being surrounded by circular feather disks resembling oversized horn-rimmed glasses perched on a schoolboy's nose. They are tiny owls and so secure in their camouflage that perched birds are prone to "freeze" and remain thus even when approached to within a dozen feet or less. Although they breed and are probably resident in the Black Hills, many saw-whets from farther north move out of their breeding grounds and drift south in winter, rarely to as far as Oklahoma. Small spruce trees seem to be a favorite winter roost site, and it is always a special thrill to peer into a snow-covered spruce and see a tiny owl intently staring back at you.

Saw-whet owls as well as the larger owls of the Black Hills coniferous forest, such as the barred owl and great horned owls, are primarily rodent-eaters. The ground-dwelling chipmunks and the more arboreal tree squirrels are part of their prey, as are various smaller mice and voles. The least chipmunk and Colorado chipmunk (Figure 53) occupy somewhat overlapping ranges in some parts of their combined ranges, as in Colorado, and are quite similar in appearance. The least is substantially smaller, roughly two-thirds the latter's size. The Colorado chipmunk is also somewhat brighter and more contrasting in its attractively striped pelage pattern, but has a relatively shorter tail, which it is prone to swing slowly from side to side rather than flicking it up and down when excited or threatened. Both are especially fond of conifer seeds, but the least chipmunk also eats many seeds of grasses and forbs, whereas the Colorado chipmunk is prone to select berries and seeds of woody shrubs and juniper seeds. For both, rocky sites for escape and breeding may be more important than the simple presence of conifers.

Chipmunks are solitary and distinctly territorial, something like typical tree squirrels. Rocky and sunny slopes are the Colorado chipmunk's favored substrate, whereas the least chipmunk is more often found in shady environments. In contrast to tree squirrels but like ground squirrels, both chipmunk species hibernate for much of the year, the least for four to seven months, and the Colorado probably less, considering the shorter winters to which this species is usually exposed. The longer summer may also allow the Colorado chipmunk to rear two litters instead of one, but in the hottest weather they may become quite inactive. Both of these chipmunks utter well-known and repeated "chip" calls, with the chip rate noticeably faster in the least than in the Colorado.

Red and rock squirrels are two similar-appearing squirrels that might seemingly be imagined as ecological counterparts in the Black Hills and Black Mesa, respectively, but the red squirrel is part of an assemblage of true tree squirrels, and the rock squirrel is part of the quite different and very large ground squirrel group. Red squirrels always nest in trees, but rock squirrels only rarely do so. Red squirrels are distinctly territorial and solitary, as are other species of their genus as well as many ground squirrels, whereas rock squirrels are relatively colonial. Altogether, their behavioral and ecological differences are greater than their similarities.

Red squirrels primarily eat the seeds of conifers, especially pines in the Black Hills, and are sometimes appropriately called "pine squirrels." Rock squirrels eat not only pinyon seeds but also acorns, juniper berries, fruits, and insects. They have even been reported to kill and eat small birds such as American robins and turkey poults. The rock squirrel is roughly three times as large as the red squirrel and is closer in both size and pelage coloration to eastern gray squirrels. Both species are noisy, and the red squirrel is likely to call almost incessantly when proclaiming a territory or confronting an intruder. Red squirrels assiduously gather pine cones and buds, sometimes removing as much as two-thirds of the pine's yearly cone crop. They strip the seeds from the cones and discard the empty cones in large rounded piles called middens. These piles are not wasted, as additional unharvested cones are typically stored within them, keeping them cool and moist enough to retain their seeds until they too can be harvested. The importance of such conspicuously exposed middens to the survival of red squirrels may be one reason for their territorial behavior. Storage of food supplies in hidden caches may make it more practical to enjoy some of the benefits of sociality.

One of the most appealing aspects of places like the Black Hills and Black Mesa for ornithologists is the certainty that at least some of the birds seen there will be ones that have a long evolutionary association with conifers, and are thus essentially unique to coniferous forests. Among them are red crossbills, pinyon jays, Townsend's solitaires, and Lewis's woodpeckers. The first time I saw a Lewis's woodpecker was when I was a young teenager on a family trip to Yellowstone National Park. I saw what I thought was a miniature crow, but it was flying in an undulating manner like a woodpecker. This was before I owned a field guide, and I puzzled over the sighting for several years, finally realizing it must have been a Lewis's woodpecker. Even today, this woodpecker doesn't seem to "belong" with the other woodpeckers, and I have little sense of why it should be nearly black. My only guess is that, like the three-toed and black-backed woodpeckers, it often is seen in recently burned areas of coniferous forest, and perhaps there black is the best color to be for reasons of camouflage. The species was discovered by the Lewis and Clark expedition, which shot several "black woodpeckers." One of these specimens was later used by Alexander Wilson to name the species in honor of Meriwether Lewis.

Since my first encounter, I have seen only a few Lewis's woodpeckers and always hope to find them on trips to the western mountains. In the Great Plains they are essentially limited as breeders to the Black Hills, but there are also a very few breeding records for northwestern Nebraska's Pine Ridge. During breeding they are especially likely to be seen in burned areas, as they seek out dead snags for nest sites. During summer this species also spends a great deal of time "flycatching" insects during slow, almost soaring flight. Like the red-headed woodpecker and northern flicker, which also tend to forage on seasonally available insects, the Lewis's woodpecker is somewhat migratory. On migration and during winter, it is more likely to be found around riparian woodlands, oak woods, or old nut orchards.

On a recent autumn trip to the Black Hills, Townsend's solitaires seemed to be almost everywhere. They perched at the very tops of tall ponderosa pines in full view, singing as if it were spring rather than fall, with a nearly continuous train of sound. The song reminded me of an American robin, but singing at almost double speed, and the males barely bothered to catch their breath between song phrases. It was practically the only bird sound to be heard in the woods, save for the steady yanking of nuthatches.

Persons unfamiliar with solitaires often find it hard to believe that they are seeing a species of thrush. They are slimmer-bodied and more streamlined than other thrushes. Unlike many of the best thrush songsters, solitaires are not inclined to hide in forest undergrowth, where catching a glimpse of the birds seems almost impossible before they vanish from sight, leaving only their increasingly distant songs to add to the frustration. When a Townsend's solitaire takes off, it is even more beautiful than when perched, as it then flashes buffy cinnamon wing patches both above and below as well as white outer tail stripes. In spite of its predilection for tall pine trees during the breeding season, the species nests on the ground, often tucked among the roots at the base of a pine or well hidden under an overhanging earthen bank.

During fall and winter, Townsend's solitaires gradually leave the Black Hills, and at least some gravitate to the upper Platte Valley of western Nebraska. There they gather on steep slopes that are covered by groves of western red cedar and seem to spend the winter months foraging mostly on cedar berries.

Of the bird species mentioned here, only the pinyon jay occurs both in the Black Hills and the Black Mesa. In the Black Hills it is found only at lower elevations in the southern hills, occurring where the trees are small and scattered, in dry woodlands rather than forestlike situations. There the birds nest in low pines, typically on the lowest horizontal limbs from six to twenty feet above ground. Likewise in the Black Mesa country, the birds inhabit pinyon-juniper woodlands, foraging mostly on pine seeds, juniper berries, acorns, and insects. Nests there have been found from eight to twelve feet above ground in junipers and pinyon pines. The birds are highly social at all times, forming noisy flocks that remind one more of crows than jays. Although the young are certainly fed largely on insects such as grasshoppers in their early life, it has been reported that the juveniles learn how to extract the pinyon nuts from the cones even before they leave the nest. Whether this is true or not, seeds of tender cones of pinyon pines and, to a lesser extent, ponderosa pines are the species' primary food resource.

Pinyon jays are highly social, with groups roosting together during nest-building, feeding socially while posting "sentry" birds as lookouts, and using "helpers" at the nest to assist in the feeding of nestlings, which include not only immatures but also adults other than the breeding pair. Feeding by helper individuals is most common after the fledglings have gathered in loose crèches, each chick screaming to be fed according to how hungry it is. Their annual nesting success is highly dependent on a reliable source of pinyon seeds, and if the seed crop is poor in one area, the birds may migrate to other areas up to several hundred miles away. The birds gather and cache stores of pinyon pine as well as ponderosa pine seeds, and such food supplies allow them to begin nesting early in the spring, while temperatures are still cold and insects are inactive. Like other corvids, they have a remarkable ability to remember the locations of such hidden foods.

Like the pinyon jays, red crossbills are also never found far from conifers. Adult crossbills seem, at first glance, to have badly malformed beaks. Their sharp tips don't meet directly, but instead cross over. Young birds, such as newly fledged ones, lack this feature and have rather typical finchlike beaks designed for handling and crushing seeds. Yet, as the beaks of crossbills mature, they become increasingly efficient tools for inserting the tips between the woody bracts of a pine cone and, by twisting the head and beak, forcing the bract open sufficiently to allow the seed to fall out. By then the crossbills have become such seed specialists that both the timing and success of their reproduction become dependent upon the size and timing of the local conifer seed crop. In various parts of North America, as many as nine discrete populations of red crossbills have evolved, each specialized for extracting local food resources. Current field research has suggested that many of these local types may actually represent separate, very similar, but distinct "sibling" species, with varied and distinctive vocalizations.

Red crossbills in the Black Hills seem to nest there only irregularly, perhaps depending on the size of the seed crop. Females in breeding condition have been collected there in December, and nests with hatched young have been seen as early in the year as mid-February. Thus, at least in certain years, nesting must begin during some of the coldest weather of winter, probably depending on the supply of available seeds. Dependent young have been observed as late as early July, suggesting a remarkably long or flexible six-month breeding cycle. The birds are not only highly social but also remarkably tame, allowing one to observe their interesting foraging and nesting behavior at close range.

CHAPTER 9

Waders, Dabblers, and Divers

The Prairie Wetlands

The gifts of glaciers past are sometimes difficult to perceive, even when their evidence is all around us. I grew up in the Red River Valley of eastern North Dakota, a region so flat that, like the ocean, one could almost imagine the earth's curvature by noticing the way that tall structures such as elevators seemed to recede into the ground when viewed from a distance of several miles. Yet, of all the people who lived in Christine, the little town of my childhood, perhaps none realized that the deep, black soil they assiduously cultivated was composed of the clay-rich sediments of a Pleistocene lake bottom, and that perhaps 20,000 years ago the area would have been near the middle of a cold-water lake so vast that its wooded shorelines were likely to have been beyond view.

My grandparents' homestead farm, seventeen miles to the west of Christine, lay on a sandy delta of a river that once poured great quantities of glacial meltwaters into Lake Agassiz's western edges. Even farther west, the "prairie potholes" that I loved and where as a teenager I first photographed waterfowl were simply the product of undulating glacial moraines where water seasonally filled the low-lying depressions, creating the greatest waterfowl breeding habitat anywhere south of Canada. None of that fascinating history was ever mentioned by the dour minister of our town's Norwegian Lutheran church, the only church in Christine. Until its steeple was hit by lightning and destroyed, the church was probably the tallest building in town, dwarfed only by the towering grain elevator beside the railroad tracks. From the church's bell tower, one could see for miles in all directions across a completely flat lakebed that had once been covered by tallgrass prairies, but by the 1930s had been planted almost entirely with spring wheat that throve in the legacy of the fertile, glacier-enriched soil. Yet, our black-frocked preacher was much more preoccupied with the prospect of forthcoming damnation and hellfire than saying, or even learning, anything of long-past glacial history or our collective human debt to it. Indeed, he would certainly have scoffed at the heretical notion that the events I have just described were written on the landscape as plainly as were the printed words on the pages of his enormous Bible.

Luckily, I eventually escaped Christine and increasingly discovered the wonders of prairie wetlands. More than any other single factor, it was the time I spent crouching in cattails and phragmites beds along the edges of just-thawing marshes, with uncountable myriads of migrating waterfowl swirling above and all around me, that shaped my entire future.

Wetlands are among the rarest of all the Great Plains habitats. Excluding reservoirs, the natural lakes, marshes, temporary ponds, and streams of the region make up far less than 1 percent of its surface area. In my *Birds of the Great Plains,* I determined that, in spite of the very tiny area represented by Great Plains wetlands, less than 1 percent, they support about 22 percent of the region's entire breeding avifauna. In that sense, wetlands probably are fifty or so times more valuable in promoting breeding bird species diversity than the overall average for all terrestrial habitats.

Second to wetlands, woodlands and forests are next most valuable. These habitats support 51 percent of the region's breeding birds, but occupy only about 15 percent of the total land area, making such habitats over three times more valuable in promoting avian species diversity than the collective average for all land areas. Grasslands at least historically covered some 81 percent of the region, but account for only 11 percent of its breeding bird diversity. This low avifaunal diversity among prairie birds is perhaps in part a reflection of the fairly recent evolution of the grassland ecosystem as compared with forests and wetlands, but must also reflect average abundance and diversity of food resources, especially during the breeding season. The relatively uniform stature of grassland vegetation, with far less three-dimensional complexity than is present in woodlands and forests, may also reduce the number of available foraging niches.

Water is of course even flatter, but it uniquely offers above-water, surface-level, and below-water foraging, and the shoreline vegetation of wetlands is typically diverse and highly productive. Of sixty-seven wildlife species (birds, mammals, and herps) identified by Deborah Finch (1992) as being of special regional conservation concern in the western Great Plains and Rocky Mountains, 55 percent are at least in part associated with wetland and riparian habitats. These same habitats support a majority of species of all these major vertebrate groups except for reptiles, which are more closely associated with grasslands. Clearly, water-dependent habitats are among both our region's rarest and most valuable wildlife habitats.

Beyond the simple presence of water, the concentrations and types of its dissolved nutrients are of great importance for both plants and animals. Some northern wetlands have acidic waters that permit only a few acid-tolerant plants such as sphagnum moss to grow. These waters are notable for their unavailability of nitrogen to plants and tend to accumulate undecayed organic matter in the form of peat. In such places carnivorous plants often occur that trap insects and extract their proteins as a means of obtaining nitrogen. At the other extreme are hyperalkaline wetlands, with dense concentrations of minerals such as sodium and potassium carbonates and bicarbonates. Only a very few organisms can survive these briny conditions, such as brine flies, brine shrimp, and some other crustaceans. Nevertheless, these wetlands are often favored foraging sites for shorebirds and some waterfowl that manage to gather such foods without becoming poisoned by the water.

Probably the most productive of all Great Plains wetlands are the only moderately alkaline marshes such as those found in the glacial moraine "prairie pothole country" east of the Missouri River in the two Dakotas. In South Dakota alone, there were still an estimated 900,000 wetlands in the late 1980s, and perhaps 1.5 million acres of wetland habitat. Here, a half dozen or more species of ducks may be nesting simultaneously, as well as four or five grebes and several elegant shorebirds. Red-winged and yellow-headed blackbirds patrol the marshy edges, marsh wrens and sedge wrens chant from hidden grasses, and black or Forster's terns hover overhead. Frog choruses add to the noise, and sometimes northern harriers slip by, unheard and almost unseen, always on the lookout for unwary voles. Such are the times when one is willing to forget about mountains, arctic tundra, or even tropical cloud forests and revel in the sheer pleasure of it all.

Some wetlands, because of water action that causes sandy bars or islands to form periodically or because of fluctuating water levels that expose sandy substrates, lack the usual shoreline border of cattails, bulrushes, and sedges typical of prairie marshes. The piping plover and least tern (Figure 54) are shorebirds that will nest only on such barren stretches of sand or gravel near water. At one time, both species were widespread along the upper Missouri River as well as many of its tributary streams, such as the Platte and Arkansas. Their breeding ecology evolved in conjunction with the historic seasonal flow cycles of these rivers. The Great Plains rivers are mostly fed by late-winter and spring snowmelt from the eastern slopes of the Rocky Mountains, plus the spring rains that are the usual precipitation pattern for this region. As a result, the rivers of the plains tended to rise and often flood in midspring, then slowly decline through the summer months. This kind of seasonal fluctuation meant that by May the rivers were usually starting to recede from peak flows, exposing sandy islands and bars. Least terns and piping plovers could then nest on these islands, fairly secure from future flooding and from at least some of the land-based predators that might attack their eggs or chicks.

This strategy worked well until technology began to alter river flows, often either entirely drying them up through excessive water withdrawals or modifying annual flow rates to control spring flooding or facilitate summer barge traffic. As a result, both species began to experience widespread population declines, and eventually both became candidates for federal threatened species status. In Nebraska, piping plovers now nest mostly along the Platte, Niobrara, and unchanneled sections of the Missouri Rivers; in South Dakota, along the Missouri River; and in Kansas, very locally along the Kansas River. Very similar breeding distribution patterns exist for the least tern in these states, but they tend to be slightly more widespread.

Snowy plovers are essentially slightly faded versions of piping plovers, and their breeding ranges practically meet in Kansas. The piping plover nests locally south to northeastern Kansas, and the snowy plover breeds north to central Kansas, specifically Quivira National Wildlife Refuge and Cheyenne Bottoms State Wildlife Area. The Great Salt Plains National Wildlife Refuge of Oklahoma offers an even better

Figure 54. Least tern, adult and chick

nesting habitat, as the birds prefer sandy beaches, shorelines, and alkali flats with essentially no standing vegetation. Somehow the pale plumage of the bird, including its downy chicks, can blend with the substrate and make them maddeningly hard to find. But this is good for the plovers, which are far too small to defend themselves from even the tiniest predators.

This relative invisibility strategy is also used by the piping plover, which I have often observed along the sandy shorelines of Lake McConaughy in western Nebraska. Not only are their eggs nearly impossible to see in the sand-and-gravel substrate they use, but their chicks are equally cryptic. Sometimes the birds make the mistake of taking their young to a large parking area along the beach, perhaps for the shade afforded by the cars. However, it is then a hair-raising experience to watch the babies narrowly escape being run over as cars come and go. Still, I have yet to see one actually be killed during these highly risky activities. Instead they have managed to flee at the last possible moment. On a few occasions I have tried to catch these little sprites in order to move them out of harm's way. However, they can run amazingly fast on their inch-long legs and can easily make a fool of a six-foot adult.

On those occasions when I have watched the nesting activities of least terns and piping plovers along the shoreline of Lake McConaughy in western Nebraska, I often have simultaneously been watched by flotillas of western grebes floating at a safe distance offshore. Western grebes (Figure 55) exude a certain air of royalty about them; their necks are unbelievably long and gracefully arched, their eyes are as red as the best Burmese rubies, and on their crowns they wear black caps that during spring are usually cocked up in a jaunty pattern somewhat resembling a three-cornered cap. They move through the water with the silent assurance of all grebes, ready to disappear below the surface in an instant should danger threaten. They are almost always found on large, shallow, and clear marshes mostly rimmed by cattails and rushes, but with open areas suitable for unobstructed fishing. They often dive with a forward jump that sends them quickly underwater, where they resemble slim attack submarines, able to outswim small minnows and easily catch them with their rapier-like beaks.

Although usually silent, during spring the marshes of the Dakotas and Nebraska fairly ring with the tinkling calls of western grebes, resembling sleighbells in the distance. Then, if one is lucky, one might see a pair of courting grebes floating easily on a sky-blue marsh lined with golden rushes, alternately dipping their bills and then quickly preening their back feathers in perfect synchrony, in a ballet scene

Figure 55. Western grebe, adult pair with chick

Figure 56. Ruddy duck, courting adults

never to be forgotten. No other waterbird quite so perfectly captures the beauty of prairie wetlands or the very essence of wildness.

I never imagined I would ever hold a live western grebe in my hands, but on an ornithology field trip to the Nebraska Sandhills I once saw one huddled on the shoreline of a large marsh, something no healthy grebe would ever do. One of my students quickly ran down and picked it up before it could reach deep water, whereupon I saw that one of its feet had been neatly amputated, most probably by the jaws of a snapping turtle. I looked at the bird with a mixture of deep sadness and total admiration for its utter beauty. Unwilling to kill it in order to preserve it as a study skin, I simply released it, thinking that I would prefer not to know how it met its end than to have its death on my own conscience.

If the western grebe is considered the last word in aquatic stream-lining, the ruddy duck (Figure 56) certainly falls toward the rear of the pack. Its primary charm lies in its rather rotund shape, including a neck that at times seems larger than its head, legs and feet that are placed so far back on its body that, grebelike, it has a difficult time standing on dry land, and a tail seemingly longer and with more spiky feathers than might be needed by any duck in the world. But the ruddy duck is the product of an evolution that has favored diving over flying and swimming over walking. The long tail is associated with underwater maneuvering, the rear-positioned legs and large feet make for effective and rapid diving, and the large neck of males is associated with sexual advertising display. The male ruddy duck is the only North American duck with a special tracheal air sac that can be inflated like a small balloon for courtship display, after which the lower bill is rapidly and repeatedly tapped downward on the neck. This action produces a soft drumming sound and forces air out of the breast feathers so that a ring of small bubbles is formed in the water around the breast. This strange "bubbling" display is made even more bizarre by the male's contrasting bright blue bill and white cheeks and his erection of two small feathered "horns" on his black crown, introducing a slightly diabolical aspect into an otherwise generally comical sight.

Ruddy ducks, like western grebes, are classic birds of prairie marshes. Both build their nests on semifloating platforms of bulrushes and other emergent plants, adding new materials as the earlier ones gradually become waterlogged and start to sink. These barely seaworthy platforms are probably safer places for a clutch of eggs than shoreline nests would be, but they tend to tip or flood should water levels fluctuate rapidly and also are susceptible to loss of eggs through wave action. Nesting success in such nests is often rather low, and ruddy ducks are not the best of mothers, even if they should manage to hatch their eggs. Perhaps it is more accurate to say that ruddy ducklings are rather delinquent offspring; they tend to stray from their mother's vicinity at a young age and may end up lost or perhaps even eaten by a snapping turtle or a low-flying northern harrier.

It is always a pleasure to see a northern harrier (Figure 57) coursing low over a wet meadow or prairie marsh. They are so graceful in their leisurely flight, and so quick to drop down on a small rodent, that they are clearly masters of their domain. They are even more exciting to watch in early spring, when males mark out their territories by flying in a long series of steep climbs and sharp dives so that a near-looping pattern is produced. The generic name for harriers, *Circus*, means circular and refers to this rather acrobatic tendency. The English name "harrier" does not refer to any possible preference for

161

Figure 57. Northern harrier, adult male (above, facial disk
aggressively ruffled) and female (below, facial disk slimmed).

hares as prey, but rather to the determined manner in which the birds
might harry their prey until it is finally captured.

Other than the American kestrel, the northern harrier is the only
raptor of the plains in which the sexes are differently plumaged. Not
surprisingly, the females and immature males are a dead-grass brown,
which closely matches the color of their grassy nest sites. Males are
mostly a ghostlike silvery gray, except for black-tipped wings and a
white rump patch. This latter trait is also shared by females. Like most
other raptors, the females are also somewhat larger than males, but in
harriers this difference is hardly noticeable in the field. Males are typ-

ically monogamous, but there have been several cases of bigamy detected in this species. In such cases, the females nest in different parts of their mate's large home range, and the male is kept busy in trying to capture enough food to provide for all of his rapidly growing young.

Northern harriers nest in wet marshes, their eggs often being situated only a few inches above the water level. The same is true of the Wilson's phalarope. Like the other two phalarope species, the Wilson's has "reversed" sexual dimorphism, meaning that females are larger, more aggressive, and seasonally more colorful than the males. They also establish breeding territories, court the duller males, and of course lay the eggs. Otherwise, it falls to the male to complete the job of incubating the eggs and rearing the young. No other North American birds have such clear-cut sexual and plumage reversals, and very few other birds of the world share this trait. One presumed potential value of such behavior is that, given a long enough breeding season, the female might sequentially mate with and lay clutches for several successive male mates to incubate and rear. Two of the three total phalarope species breed in the Arctic, where the nesting season may be too short to make such behavior practical. The Wilson's phalarope nests south in the Great Plains regularly to Nebraska and perhaps rarely or locally in Kansas, where the breeding season is appreciably longer. Thus, if polyandry is to occur at all, it should most often occur here. Local distribution of resources may also affect the incidence of phalarope polyandry, as has been reported for the spotted sandpiper.

In the Nebraska Sandhills, the Wilson's phalarope is a species that is so closely associated with highly alkaline wetlands and their attendant fauna of brine flies and other brine-adapted invertebrates that one can predict with absolute certainty just what ponds and marshes are going to support breeding phalaropes. There too American avocets might sometimes be found foraging as well as the occasional black-necked stilt, but these species are not nearly so closely associated with the brine ecosystem as are phalaropes. Although phalaropes are the most aquatic of typical shorebirds, their webbed toes allow for foraging by swimming rather than just wading, and it is always a pleasure to see a group of them busily swimming while rotating in place like tops, this circular movement stirring up the water below and bringing very small aquatic organisms to the water surface. At times phalaropes will also run along the shoreline, grabbing newly hatched brine flies as quickly as they emerge from their pupal cases.

Another wetland-nesting species that reaches the peak of its abundance on the northern prairie marshes is the Franklin's gull. It is this species that one often sees in spring flocks over the prairie, the black-capped birds following excitedly and landing immediately behind farmers engaged in spring plowing or tilling. Thereby they can pick up insects exposed by the farm implements and easily eat their fill. Such flocks are usually on their way to breeding marshes in the Dakotas, northwestern Minnesota, or perhaps southern Canada. Great nesting colonies sometimes form on large prairie marshes, with pairs occasionally numbering in the thousands or rarely even the hundreds of thousands. A few scattered accounts of breeding in Kansas and Nebraska have been reported, but these are isolated and probably atypical events. It is true that nonbreeders regularly oversummer in some marshy areas of the central plains, but they should not be mistaken for breeding adults.

Ring-billed gulls and California gulls now also breed from South Dakota northward, as do common and Forster's terns. The insect-eating black tern nests commonly south to the Nebraska Sandhills and very rarely to central Kansas. These are all visually attractive birds in their own way, especially the terns, but only the Franklin's gull is usually considered a Great Plains endemic, and it is generally regarded as the most attractive of our breeding gulls. A traditional North Dakota name for it was "prairie pigeon" or "prairie dove," both of which seem to express perfectly its grace and beauty in flight.

Franklin's gulls build their nests over water in emergent marsh vegetation, in a manner similar to those of the western grebe and ruddy duck. They are no less vulnerable to flooding, drying out, tipping over, or being destroyed by waves than these other species, and their eggs or nestlings also often fall prey to crows, raccoons, minks, and aquatic snakes, all sources of mortality that can occur in a prairie marsh.

Even more abundant than gulls, grebes, or ducks on prairie marshes are the blackbirds. Although somewhat similar in size and appearance, the red-winged blackbird and yellow-headed blackbird (Figure 58) provide some interesting comparisons. The yellow-headed blackbird is essentially confined to the larger and deeper marshes of the Great Plains and American West, where the birds nest colonially in emergent vegetation. They reach their maximum densities in the glaciated potholes region of the Dakotas and are rare east of Minnesota. The red-winged blackbird, although often a colonial nester in

Figure 58. Yellow-headed blackbird, adult male

marshes, is much more tolerant of smaller, shallower wetlands and may even nest in roadside ditches with little or no water or in other almost upland situations. As a result, it nests nationwide, but reaches its densest concentrations in the far American West and the Southwest. Its total nationwide breeding population is unknowable, but at least from North Dakota through Kansas it is in the top five species as to overall breeding ubiquity or commonness, and in Oklahoma there are breeding records for nearly every county.

One national estimate was of 190 million birds in the 1970s, making this perhaps the most common of all North American birds; such are the advantages of broad ecological tolerance. During fall and winter millions of red-winged blackbirds migrate through the Great Plains states, with concentrations in some single refuges occasionally exceeding a million birds. For example, during one Christmas count in 1965, 40 million red-winged blackbirds were estimated to be present at a single Arkansas site. Nationally the species is declining at a slight but significant rate, whereas the population trend for the yellow-headed blackbird is a slightly positive one. Humans who consider these blackbirds a major pest should recognize that we have put out a gigantic feast for them in the form of vast grainfields, and they have simply taken advantage of it.

Just as we associate ruddy ducks, western grebes, and Franklin's gulls with the prairies, we are inclined to think of beavers only in association with the northern and western American forests, where they provided a major impetus for the early exploration and exploitation of North America's rich resources. Yet, they have also traveled out into the Great Plains grasslands along our river systems far from any real forests; I have seen creeks so narrow one could jump across them that have had the unmistakable signs of beaver work along their edges in the form of lopped-off saplings and trees and efficient dams constructed of logs and branches. More than any other wild mammals, beavers are able to modify their environment in a way that benefits them. These changes in the habitat also allow a wide variety of water-dependent animals including fishes, amphibians, reptiles, birds, other mammals, and invertebrates to survive and flourish.

The dams that beavers construct are truly engineering marvels. I have seen beaver dams in the Rocky Mountains constructed with such beautiful inward curves one would think that beavers discovered the principles of arch design and construction long before the Romans began building viaducts. Indeed, one species of beaver is native to Europe, but it is not so well known for its dam-building abilities as is the American species. Beavers are North America's largest rodents, with males often exceeding sixty pounds and known even to rarely reach ninety pounds. Such large animals are tremendously strong; the apparent record for the largest tree ever felled by a beaver was a cottonwood 110 feet high and over five feet in diameter! A tree five inches in diameter can usually be felled in only about three minutes. Typically, a tree is cut down by a single animal, but on large trees two or three may work cooperatively. Often a small tree is cut down every other night, which is then chewed into smaller sections up to about eight feet in length. These branches and their leaves are then hauled away, to be incorporated into a dam or lodge or stashed into a deep pool near the lodge by jamming one end of the branch firmly into the muddy bottom. Eventually such cached items will be recovered, and their bark and twigs consumed as food. About a third of the cellulose that is consumed eventually gets digested.

Beavers are among the few rodents with very strong pair- and family bonds, a trait that is probably needed because of the degree of cooperation required for the physical labor and complex engineering feats that are part of this species' survival strategy. It is thought that beavers pair indefinitely, perhaps even for life when both are about the same age. It is known that, unlike muskrats, there is only a single litter per season, with three or four young typical. These youngsters have a long weaning period of about six weeks and, unlike muskrats, remain for an extended period within their parents' lodge, probably in most cases until they are two years old. As a result, the usual colony size is about six animals, representing three generations. Most beavers do not mate and breed until they are approaching three years of age, but this may depend on opportunities for local breeding or dispersal. Because of their size, beavers have few predators to contend with and may live in the wild for up to about fifteen years. Beaver densities in the relatively unforested Great Plains are of course quite low, but in prime historic habitats might have reached as many as fifty per square mile. Probably only in national parks and preserves are high densities reached, as the efficiency of beavers in building dams and excavating underground tunnels regularly brings them into conflict with human wishes.

If beavers can be regarded as the engineers of the American wetlands, then muskrats perhaps deserve the title of city planners. In the open plains, where trees are few and far between, the muskrat actually takes over both roles, constructing not only thoroughfares through heavy marsh vegetation but also "houses" of marsh weeds, roots, and branches. These large constructions, up to about eight feet in diameter and four feet high, not only provide safe homes for a muskrat family but also offer sunning spots for turtles and perching places or even nest sites for birds such as ducks, geese, and other waterbirds. Canals leading from the house to foraging sites are dug out, making easy swimming routes for aquatic birds. Nests of species such as grebes, ruddy ducks, and canvasbacks are often located close to such escape paths, sometimes on the feeding and resting platforms already constructed by the muskrats.

The houses of muskrats are used year-round, but especially during breeding. Their thick walls provide some insulation from winter cold as well as summer heat, and the central chamber may be subdivided to provide semiprivate living quarters. They are typically occupied by a breeding pair, but during winter several animals might share a single den or house. The breeding season begins in late winter or early spring, and in the central plains states there are often three mating peaks per year, in March, April, and May. The gestation period averages four weeks. Because mating occurs immediately after the female gives birth, a succession of litters is produced in a very short time, namely about once a month. Generally four to seven young are produced per litter, and weaning occurs at only three or four weeks of age, so that a single generation of dependent young should be present in the den at any time. Often the older youngsters are moved to new living quarters as the next litter is being raised, perhaps in part to prevent them from attacking and eating their younger brethren.

By six months of age, the young are adult-sized and ready to breed, but it is likely that few if any breed during the year in which they were born and instead must wait until the following spring. With such a high birth rate, it is obvious that the mortality rates of muskrats must be correspondingly high. Probably only about a third of the young produced each summer survive until their first winter. Weasels, minks, raccoons, snakes, coyotes, and a host of other predators may locally rely on muskrats for their food supplies, and others may die from drowning, fighting, disease, starvation, or miscellaneous factors. In favorable habitats, their densities may reach ten to twenty animals per acre.

Where shoreline banks are fairly steep, the animals usually prefer to make bank dens rather than construct houses. These are simply long tubular excavations, usually with underwater entrances, that extend back from the shoreline as far as fifty feet and with diameters of about six inches. Where aeration is a problem, an air shaft to the surface might be constructed. Such excavations are often built into the walls of earthen dams that form small farm impoundments, and in such cases the actions of muskrats, like those of beavers, may come into conflict with human intentions.

The working of beavers and muskrats makes life possible, or at least easier, for a wide variety of other water-dependent wildlife species, such as frogs and aquatic turtles and snakes. The Plains leopard frog (Figure 59) is the common pond frog species that is familiar to almost every child of the Great Plains, or at least those living south of the Platte River. The Plains leopard frog is gradually replaced farther north by the very similar northern leopard frog (Figure 59), which for all intents and purposes is essentially identical in appearance. Where they occur together in northern Nebraska and southeastern South Dakota, the Plains leopard frog is generally recognizable by being more tan dorsally, whereas the northern leopard frog is more greenish, with its dark spots more clearly bordered by whitish coloring. In the Plains leopard frog, the white line extending laterally from the snout down the upper trunk to the tail is typically broken at the base of the hind leg, but is continuous in the northern leopard frog. The Plains leopard frog also has a pale yellowish spot on the eardrum, or tympanum, and a more prominent yellowish jaw stripe. The mating call of the Plains leopard frog is a series of two or three guttural notes per second, producing a chuckle-like sound, whereas that of the northern leopard frog is more snorelike, lasting about three seconds.

Unlike many other amphibians, the leopard frogs are active over a long period, from about February to October in the central Great Plains and even occasionally in the winter. Breeding may occur as early as February in Kansas and Oklahoma, but usually starts during April in the Dakotas, as soon as ponds and streams become ice-free. They are perhaps the commonest native amphibians and as such are often captured to use as fish bait or for other purposes. In Nebraska, nearly 175,000 northern leopard frogs and about 14,000 Plains leopard frogs were reported sold in or exported from Nebraska during the period 1994–2000, a figure that excludes all those taken for noncommercial use. Many of these were sold to biological supply houses for educational use.

A much rarer herptile in the Great Plains is the Blanding's turtle (Figure 59), which reaches the western edge of its range in the wetlands of the Nebraska Sandhills. It is a relatively aquatic turtle, favoring permanent marshes, and is particularly appealing to humans because, dolphinlike, its mouth profile has a built-in smile. Although

Figure 59. Blanding's turtle (top), Plains garter snake (middle), and northern (lower left) and Plains (lower right) leopard frogs. The eggs and tadpole stage of the Plains leopard frog are also shown.

its lower shell is hinged as in box turtles, it cannot close its shell completely, and its upper shell (carapace) is not so strongly arched as in the box turtle. It has an unusually long neck, with which it can snap out and grab insects, crayfish, and other live foods. Some plants are also consumed. Like the box turtle, it makes a dug-out nest in which six to eleven eggs are typically deposited.

Nearly all (95 percent) of the Great Plains snake species are harmless to humans, although any is likely to bite if caught and picked up. The Plains garter snake (Figure 59) is one of the most common snakes in the region. This snake is named for the decorative garterlike linear pattern that runs along the sides and top of the body. The body of adults is mostly gray to tan-colored on the sides, interrupted by a linear series of small back spots, and with a middorsal stripe of bright yellow orange as well as a pair of lateral stripes that are usually also yellow or greenish yellow but sometimes bright red. They are medium-sized snakes, up to a maximum of three feet long, with females slightly larger than males, but males having longer tails than females.

Grassy prairies and the edges of wetlands are the favorite habitats of Plains garter snakes, where they are common and feed on almost anything they can catch and swallow, up to good-sized toads, frogs, and rodents. Frogs are perhaps their most common prey. They lack the heat-sensing abilities of pit vipers and largely rely on olfactory cues as detected by their sensitive tongues. Being nonvenomous they have only limited powers of self-defense, but when captured they emit foul-smelling material from their anal glands and try to smear it on their captors.

In the central plains, mating by copulation occurs during spring and sometimes also during fall before the animals retreat to their winter dens. The females are ovoviviparous, meaning that they produce eggs that hatch within their reproductive tracts, giving birth to their young during the latter part of the summer. As many as sixty young may be produced in a single birthing event. A number of predatory birds feed on garter snakes, as do some mammals and even other snakes.

The only toad endemic to the Great Plains region is the appropriately named Great Plains toad (Figure 60). Like other American toads it is large, slow, and seemingly content with a leisurely if not indolent approach to life. Earthworms and large insects are among its favorite prey. It is a fair-sized toad, growing up to about four inches long, with many dark greenish blotches that are outlined by cream or yellow, and

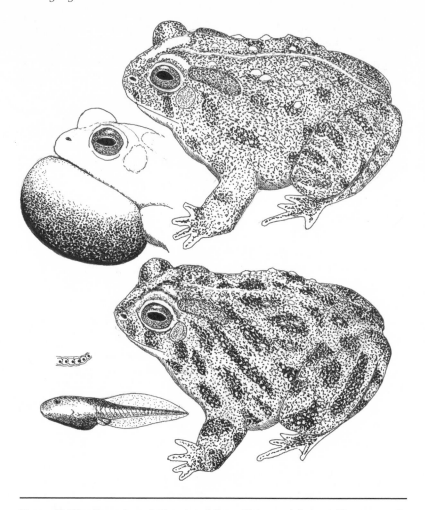

Figure 60. Woodhouse's toad (above) and Great Plains toad (below). The extent of throat inflation during male calling and the eggs and tadpole stage of the Great Plains toad are also illustrated.

the entire upper surface is covered by tiny wartlike growths. A pair of bony crests arise behind each eye, which merge to form a V and continue forward as a knob on the snout. It differs from the very similar but more western-oriented Woodhouse's toad (Figure 60) in that it is more strongly patterned, with the dark dorsal blotches larger and more clearly pale-rimmed, but the pale line down the middle of the back is not so evident. The Great Plains toad is also more often found in grassy habitats well away from water and sandy areas, but must return annually to ponds and pools for breeding. This occurs in spring and summer, the availability of surface water permitting. Like other toads, the male's throat is greatly inflated during chorusing, extending forward and reaching in front of the snout. The call is a harsh and pulsating chugging, with individual call sequences lasting thirty to fifty seconds. By comparison, the call of Woodhouse's toad is a nasal noise lasting up to about three seconds, sounding like a bleating lamb. The eggs of both species are laid and fertilized in water, and the tadpoles also must mature in water. As colder weather returns, the toads retreat underground for the winter.

Although the skin warts and large shoulder glands of toads contain toxic secretions that render them invulnerable to some predators, they are the favorite prey of hognose snakes as well as some other snakes, skunks, and various predators with seemingly limited concern about their vile taste. If grabbed, the toad will quickly swallow air and inflate its body, making it hard for the predator to swallow it. It will also hunch up, better exposing its poison glands. Yet, hognose snakes pay little attention to any of this, and their fangs are developed in such a way that they can often puncture the toad and deflate it, making it easier to swallow.

CHAPTER 10

Cowbirds, Coyotes, and Cardinals

The Wildlife around Us

It is increasingly and unfortunately true that ever more of the people who call the Great Plains their home, and perhaps were even born and raised here, have little or no conception of its natural attractions or inherent values. To many of them, a marsh might be regarded as a simple wasteland, useful only as a potential place to be drained and filled, a river something better dammed than "wasted," a prairie nothing but useless grass, and a spring migration simply a trip to the Florida or Texas coast in order to get drunk. Our break with the natural world has been so complete that we think nothing of watching television programs in which wild animals talk, every nocturnal scene from anywhere has some obligatory but often biologically inappropriate owl or loon call, and when we hear such terms as eagles, falcons, orioles, and lions, we are more likely to associate them with athletic teams than with wild animals.

Yet, even in large cities we are surrounded with wildlife, whether it be the fortunate few native species that have adapted to human environments, such as squirrels, cottontail rabbits, blue jays, and American robins, or nonnative intruders, who as tagalongs from Europe or elsewhere have learned to survive here too. These include such familiar species as European starlings, house sparrows, rock doves, house mice, and Norway rats. Most of these we have come to regard as pests, although the ring-necked pheasant, gray partridge, and a few other introduced birds and mammals have become important game species.

More often they have eventually proved to be more of a problem than a benefit, and almost invariably the most successful imports have managed to achieve their success at the expense of one or more of our native species.

Most of the Great Plains wildlife has suffered from the effects of humans, some to a very great degree. Yet, others have not only adapted but in a few cases have also thrived. To a large degree, these are "edge-adapted" species, which once mainly occurred along places of abrupt ecological shifts, such as along the borders of woods, where they encountered grasslands. Human settlements have greatly increased the amounts of available "edge" habitat by opening and fragmenting forests, planting trees in once treeless areas, building highways and railroads, digging ditches, erecting bridges or buildings, establishing farmsteads, and generally "civilizing" the land.

Many edge-adapted species of wildlife, as well as some otherwise ecologically preadapted species, quickly responded to the presence of humans, including nearly all those birds and mammals that most often are now seen around towns, villages, and farms. Thus, we have chimney swifts that nest exclusively in chimneys rather than trying to seek out hollow trees, purple martins that nest almost entirely in communal birdhouses, common nighthawks that nest on flat, gravelly rooftops, and barn owls and turkey vultures that nest in abandoned buildings rather than using natural recesses. Various gulls, mice, rats, and opossums now feed largely on urban trash, coyotes have learned that they can survive on rats and mice around farms, and red foxes have found that they can escape coyotes by living in the suburbs. American crows have similarly increasingly adapted to city life, where they live fairly free of hawks by day and owls by night. Chickadees, mourning doves, American robins, blue jays, and northern cardinals all readily exploit the largess of backyard bird feeders, as do the less appealing European starlings, common grackles, house sparrows, and brown-headed cowbirds. Eastern screech-owls survive and breed much better in the suburbs than in the country, thereby largely escaping attacks by great horned owls, and cottontail rabbits and fox squirrels have discovered much the same. Tree swallows nest in birdhouses where hollow tree sites are lacking, barn swallows regularly attach their mud-and-twig nests to barns and houses rather than seeking out rocky cliffs, and cliff swallows use bridges rather than rock faces for their dried-mud nests. Another cliff-dependent species, the peregrine

Figure 61. Peregrine falcon, adult

falcon, has found that skyscraper ledges work nearly as well as high cliffs for nesting, and additionally they have an unlimited supply of rock doves to forage upon when nesting in large cities.

Like wild animal populations, human populations on the Great Plains are also dynamic and mobile. A series of drought years, combined with low crop and livestock prices, has driven people out of the western and more arid parts of the Great Plains states. During the decade of the 1990s, large areas of the western plains were abandoned by humans, as they moved to larger cities or to other states. During that period, average population losses of up to 5 percent annually occurred in 49 percent of North Dakota's counties, 13 percent of South Dakota's and Nebraska's, 10 percent of Kansas's, 6 percent of Oklahoma's, and 8 percent of the Texas panhandle's. Average farm and ranch sizes have increased correspondingly, in an effort to cope with the changing economics, and less land has been left wholly unmanaged. However, the federal Conservation Research Program (CRP) of the late 1980s and 1990s has helped to protect some of the most erosion-prone and marginally productive lands from further destruction by the planting of now mostly native grasses. The beleaguered Endangered Species Act also has helped for the most critically rare species such as the peregrine falcon and whooping crane, although the number of seriously threatened species whose conservation attention is "warranted but precluded" because of inadequate federal funding approached 300 nationally at the turn of the twenty-first century.

One of the few bright spots in North American conservation efforts has been the peregrine falcon (Figure 61). Brought to the brink of extirpation in North America during the "hard pesticide" era of the 1940s through the 1960s, a massive federal effort combined with the work of private groups such as the Peregrine Fund has seemingly turned the tide. Only the day before writing this section, I happened to flush a peregrine from a wooded area while driving a country road within ten miles of Lincoln, Nebraska. If I had reported such a sighting about twenty years ago, my sanity would have been rightfully questioned. Under a breeding program directed by state and federal conservation agencies and carried out by volunteer efforts of Raptor Recovery Nebraska, nearly twenty peregrines have been hatched and fledged during the last decade from Woodmen Tower in Omaha, the city's tallest building. Peregrines even have tried to nest just under the gold-leafed dome of Nebraska's state capitol building in Lincoln, where they have used the nighttime lights that illuminate the tower to locate and prey on night-flying birds such as nighthawks as well as their usual quarry of daytime foragers. Although these are the only known nesting sites in Nebraska, they are the first in nearly a century, for in presettlement days the tall cliffs of Nebraska's Pine Ridge region probably provided the only ideal nesting habitats for the species.

An older name for the peregrine is the "duck hawk," and it is true that in some undeveloped areas the peregrine is a fierce predator on many medium-sized and fast-flying birds such as waterfowl. Yet, in the city skyscraper environments where many peregrines now nest, the most common prey species consist of rock doves and starlings, and having a peregrine in the neighborhood is one of the most effective ways known for keeping a city's rock dove population at a tolerable level.

Whether hunting rock doves or mallards, the peregrine is a bolt of lightning on wings. Its favorite mode of attack is to circle above possible prey at a height of a few hundred feet, sometimes "hiding in the sun" so as to reduce its chances of early detection. Then, it folds its wings and enters a steep dive, or "stoop," reaching speeds that have been hotly argued but generally asserted to approach 100 miles per hour. It attacks its flying prey with widespread toes, its rear talons typically tearing into the bird and instantly knocking it senseless. As the quarry falls to earth, the peregrine is close behind, "mantling" the prey with open wings as it alights and delivering a final blow by severing the victim's neck vertebrae with a single bite of its sharp beak. Peregrines sometimes kill prey too heavy for them to carry off, but generally they are able to lift and carry quarry weighing up to about a third of their own weight. As females are considerably larger than males, their prey-carrying abilities are better than those of males.

Like peregrines, barn owls have different nest-site tendencies now than they did in presettlement times. Then they had to depend entirely on natural cavities, such as tree hollows or recesses in vertical cliff-sides, for nest sites. With settlement came opportunities for nesting in abandoned buildings, barns, and other outbuildings associated with ranches and farms. A ranch or farm with a pair of resident barn owls is lucky indeed; a single barn owl might consume as many as a thousand mice in a year, making them much more effective than barnyard cats in controlling rodent pests. Yet their ghostlike presence, and their sometimes bloodcurdling nighttime screams, does not always make them the most beloved of country neighbors, and barn owls often are unjustly blamed for poultry losses that were more likely to have been caused by great horned owls. For these and other reasons, such as general habitat losses, the barn owl has become ever less frequently encountered in North America.

Of all our common owls, the barn owl perhaps has the finest hearing. Its rather heart-shaped face, with two nearly circular disks of feathers radiating backward from the beak, is essentially a double parabolic disk that is wonderfully adapted for binaural sound reception. Each disk of feathers directs sounds into its respective ear opening, and each of the ear openings has a flexible skin flap beside it to help further detect the sound waves. The two ear openings are asymmetrically located on the head, with one nearly an inch higher than the other. It is believed that this strange anatomical arrangement is associated with detecting the angle above or below the horizontal from which a sound source is coming. Determining the exact position of a localized sound source such as a moving mouse in the horizontal plane is highly precise, likely to be accurate to within a few degrees, or entirely sufficient to home in on an unseen rodent from as far as about thirty feet away.

Barn owls, like other owls, have such remarkably soft and delicate plumage that a human is almost compelled to reach out and try to touch it. In that respect, owls and doves have some external similarities that make them especially appealing to humans. Doves and most owls also have soft, gentle voices that are nearly hypnotic in their rhythmic and soothing effects. It is therefore somewhat surprising that the mourning dove (Figure 62) has never been selected as an official state bird. Few birds have more beautiful voices, few are more simply elegant, and few are more abundant in North America. Based on breeding bird surveys and related information, it is likely that mourning doves are the seventh-most-common breeding bird in North Dakota, second-most-common in South Dakota, in first place in both Nebraska and Kansas, and probably tied for first place in Oklahoma. Nationwide, their numbers were estimated as approaching 500 million in the early 1990s. At least in part, this remarkable abundance results from the ability of mourning doves to nest almost anywhere, including on the ground in grassland areas, in bushes where these are the most common potential sites, and in trees wherever such are available. Furthermore, although the birds consistently lay only two eggs per nesting cycle, they begin a new nest cycle almost before their last brood is fledged, so that four or five broods per breeding season might be produced in a good year. Few if any other North American birds seem to enjoy parenthood as much as mourning doves.

Although mourning doves may not win any awards for their brilliant colors, I was amazed about a year ago when I examined a number of skins of Great Plains birds under ultraviolet lighting. It is now

179

Figure 62. Mourning dove, adult male calling

known that many, perhaps most, birds can see into the ultraviolet end of the spectrum, something that mammals are evidently unable to achieve. It seems likely that birds might therefore use such ultraviolet colors for their own social signals, and I thought it might be interesting to see if bird feathers reflect light high in the ultraviolet range. I didn't expect the mourning dove to be of any interest in this regard, but nevertheless tried one. I was astonished to see that the little black "teardrop" that is artfully located on each cheek of the mourning dove, where a beauty mark might be placed on a woman's cheek, reflected with an intense purple-violet brilliance. The sides of the neck, with their slight iridescence that is only apparent when a male mourning dove calls with an inflated neck, were also much more evident. It seems that the mourning dove is much more colorful than we have thought.

Barn swallows are another of the birds that are so familiar to humans living on farms and in cities that we sometimes scarcely seem to notice them. In Nebraska and Kansas, they are second only to the mourning doves as to frequency of occurrence in breeding bird surveys. In Australia, a very near relative is called the "welcome swallow," a name I have always loved. Our birds consistently appear in Nebraska about the second week in April, just as spring is being welcomed with the first wildflowers of the year. They also remain on their nesting grounds later than the other swallows; this past autumn, a few were still flying above Nebraska farmyards in late October and even into early November, trying to gather in just a few more insects before heading south. Cliff swallows and tree swallows had already left for warmer lands in Central or South America at least a month or more previously.

It is interesting that the five common swallows of the Great Plains, the barn, cliff, tree, bank, and rough-winged, are all nearly identical in size and probably all forage on much the same insect food base. Yet, each has a somewhat different nest-site requirement, and it seems likely that these differences help to spread out the species and reduce local competition. Although the barn swallow and cliff swallow both build adobe-like nests on near-vertical surfaces, the cliff swallow makes its nests entirely of dried mud, which it can attach to only slightly irregular or even entirely smooth rock surfaces. The barn swallow is less adept and looks for sites having some solid support from below as well as from the side, also using a mixture of plant fibers for additional strength. This minor difference has meant that barn swallows have easily been able to utilize the eaves of houses, porches, and other buildings for nest sites, while cliff swallows have largely adapted to the vertical surfaces of bridges, thereby tending to make barn swallows birds of farms and cities and cliff swallows birds of highways and river valleys.

In both cases, the birds' ranges have probably spread in historic times, and they have likewise probably increased considerably in abundance. The cliff swallow seemingly increased at an annual rate of 2.5 percent during the breeding bird survey period 1966–2000, although this figure is not statistically significant. Some of its densest populations known occur on the sides and underparts of bridges over the Platte River in western Nebraska, where at times a single bridge might support several thousand nests. Where vertical cliff faces are available, the birds still nest there too, as along some of the rocky shorelines on the south side of Lake McConaughy. I once helplessly watched nest after nest, all filled with baby swallows, being wetted by an evening storm with strong winds and heavy rains. The adobe nests thus were loosened from the cliff surface and dropped by the dozens into the wave-tossed lake. The parents flew distractedly back and forth, completely unable to stop the wholesale destruction of the colony. By the following morning, the entire nesting colony had vanished.

Like barn swallows, the long, deeply forked tails of scissor-tailed flycatchers may serve a dual purpose. Not only do these feathers make effective ornaments during aerial courtship, but forked tails also seem to increase the aerodynamic abilities in slow-speed flight, an ability that is probably very useful in catching flying insects. Immatures and females also have forked tails, although their outer tail feathers are not so obviously elongated and ornate as in adult males. Adult males also have the brightest red marking on the undersides of the wings, a feature not usually evident among perched birds, but probably an important social signal.

The flycatchers comprise the largest family of songbirds in the world, and one that is limited to the New World. Most occur south of the U.S.-Mexican border, and bird identification in Latin America is not made easier by the tremendous array of flycatcher species that often surrounds one. Fortunately, the scissor-tailed flycatcher is an exception. Nobody is likely to forget the first scissor-tailed flycatcher that he or she sees, and it is one of the great pleasures of the southern Great Plains states, from central Kansas south, to watch the antics of these spectacular birds on their breeding grounds.

Few North American birds are so widely recognized, or so obviously loved, as the northern cardinal. No doubt most urban people take up backyard bird-feeding in hopes of attracting cardinals and consider satellite species such as chickadees, American robins, house finches, and all the other typical backyard birds as only second-rate. To add to the amazing beauty of the males, there is also its wonderful song, one of the cheeriest and loudest of all urban birds. It is also usually the earliest to begin singing in late winter; even by mid-January in Nebraska there is a slight chance of hearing a cardinal singing. It is not surprising that cardinals have been chosen as the representative state bird for seven states.

When I was a child in southeastern North Dakota, I never saw a cardinal; their year-round range was simply too far to the south. Yet, in the past half century, they have moved slowly but progressively northward and now breed fairly commonly in the southern parts of North Dakota's Red River valley. Perhaps this range expansion is mostly due to climatic amelioration, but in part it also may reflect the influence of bird-feeding, assuring even the northernmost of northern cardinals a reliable food source over the long winter period.

As compared with the northern cardinal, mourning dove, and barn swallow, three of the most familiar and widespread of the breeding birds of the Great Plains, the brown-headed cowbird is unlikely to make anyone's list of favorite birds. Considering the five plains states from North Dakota through Oklahoma, it is among the ten species judged most abundant or most frequently reported as breeding in all five and rises as high as third place in South Dakota. If an index value of 1.0 is used to represent the estimated most abundant breeding species for each state and 10.0 represents the tenth, the mourning dove ranks first among the five states collectively, with an average index value of 2.4. The red-winged blackbird ranks second, with an average index of 3.8. The cowbird is third, with a mean index of 5.8. The barn swallow and common grackle are both widespread and very common breeders in Nebraska and Kansas, but do not rank among the top ten in the other three states. The national population of the brown-headed cowbird has been estimated as high as perhaps 20 million to 40 million birds, but in recent years it has been declining at a rate of about 1 percent annually.

It is likely that the brown-headed cowbird has always been fairly common on the Great Plains and originally was ecologically associated with bison, not cattle. By remaining near a grazing animal's feet, the birds can easily catch any insects stirred up by the almost constantly moving beasts. Cattle are much more lethargic than bison, but now represent nearly the only game in town, and so cowbirds have acquired their current name.

If cowbirds were notable only for their association with large ungulates, they would be a welcome part of our scenery, but they are effective "brood parasites," leaving their eggs in other birds' nests. Thereby they represent a serious threat to virtually every songbird that breeds in grasslands, as well as to some forest-nesting birds that nest close enough to the periphery of woods to fall prey to the widely ranging cowbirds.

Although the common cuckoo of Europe and its near relatives have evolved an almost unbelievable array of specialized tricks to help fool their songbird hosts into accepting the cuckoo's eggs and rearing their young, the cowbird has seemingly adopted the viewpoint once expressed by Woody Allen, namely that 95 percent of success may be achieved by simply showing up. Cowbird eggs have "shown up" in the nests of more than 200 species of songbirds throughout North America, but their eggs are most frequently laid in the nests of grassland-nesting birds, the ecosystem in which cowbirds first evolved. There is no special trend toward specific egg color or pattern mimicry; instead the cowbird egg is of a generalized spotted pattern that vaguely resembles those of dozens of grassland sparrows. The egg is slightly larger than that of sparrows, and its shell is somewhat thicker, making it more resistant to breakage. Although females typically lay only a single egg in any one host nest, the abundance of cowbirds in grassland areas makes it likely that many nests will contain at least two eggs. The incubation period is not significantly shorter than those of most of its hosts, namely about eleven to twelve days, and the young show no tendency to mimic the appearance or behavior of the hosts. Instead they simply scream for food more often and more effectively than do the host young, thus getting most of the food that is delivered to the nest by the hard-pressed parents.

Some host species will accept cowbird eggs, perhaps either not recognizing them or being physically unable to pierce or remove them. Others will abandon their nest or, as with the yellow warbler, cover over their entire first clutch with grass and begin a new one directly above it. Still other species, indeed most grassland songbirds, accept and incubate the foreign eggs along with their own, attempting to raise all the young that hatch. Such host parents usually seem compelled to feed them no matter how unlike their own young the baby cowbirds might appear. It is a case of parental attachment carried to extremes, often resulting in the starvation of their own young prior to fledging.

Unlike the brown-headed cowbird, one cannot argue that the common grackle (Figure 63) isn't beautiful; a male performing his intimidation display, with his iridescent chest and neck feathers fluffed, his tail spread, and his wings slightly spread, surprisingly resembles a miniature version of some of the long-tailed and violet-hued birds-of-paradise, such as the wonderful *Astrapia* species. Being able to see such beauty in one's backyard is a reasonable recompense for the fact that grackles are noisy, boisterous birds prone to steal and eat the eggs and nestlings of other songbirds. They also gather with vast numbers of other "blackbirds" during fall migration, descending on both cities

Figure 63. Common grackle, adult male with hatchling skink

and countrysides in search of grain crops or anything else that happens to be in good supply. By late fall they typically move out of the central plains states into wintering areas mostly in the southeastern states, where during Christmas counts of the 1960s and early 1970s, they were responsible for the second-highest average overall abundance value nationally, nearly 10,000 birds seen per party-hour. Their maximum breeding densities occur to the east of the Great Plains, as they are much more attracted to woodland edge or savanna habitats than to open grasslands.

Grackles are fairly close relatives of the brown-headed cowbird, and it seems only logical that their nests should be heavily parasitized by this species. Yet, parasitism is extremely rare, apparently because of avoidance of grackles by the cowbird and a significant amount of egg rejection or nest desertion when nests are parasitized.

If grackles and cowbirds are generally disliked by most Midwesterners, coyotes, skunks, and other natural predators tend to be positively hated. Our views and attitudes about the coyote (Figure 64) have been shaped by far too many children's stories and hunters' tales of the

Figure 64. Coyote, adult

Wild West—a villain is a basic part of almost any good story, and wolves and coyotes have made admirable villains for as long as there have been good storytellers. Native Americans considered the coyote as a trickster-figure, a godlike mixture of magician and practical joker. It is a gentler view of coyotes than our nearly universal current one, a wily, cunning, wicked, and all-around unpleasant animal that is best shot on sight. I have seen too many dead coyotes strung up on barbed-wire fences, apparently serving as some sort of gruesome trophies set out by anonymous marksmen to show just who was boss.

All things considered, the numbers of rats and mice killed in a year by coyotes far outstrips their occasional depredations on poultry and livestock. In North Dakota, the costs of hiring federal predator control agents during one recent year were nearly ten times more than the estimated value of livestock losses to predators that year, to say nothing of the benefits that coyotes had performed in rodent control. We have long come to accept the fact that our government would soon collapse if it got rid of all the people on federal payrolls doing useless work, but hiring agents to poison and trap basically beneficial animals should strain one's tolerance for stupidity. Beyond that, the number of more clearly beneficial "non-target" animals killed by such efforts is unknown. Many of the eagles that have been brought into our raptor recovery center in Nebraska had been caught by traps set out for coyotes and other "vermin." If they survive at all, they are likely to be missing a toe or even a leg, making survival in the wild impossible.

In spite of more than a century of unrelenting persecution, the coyote survives. Its range and abundance actually increased after prairie wolves were eliminated, inasmuch as wolves were one of its few natural enemies. Like crows, foxes, and other intelligent omnivores, it has learned locally that it can survive in the shadows of the suburbs and even in larger cities, sometimes eating from trash cans and avoiding guns by living within city limits. At times it may eat food set out for dogs and occasionally may even interbreed with them. All of this may be something of a social comedown for the Native American's classic trickster, but in the final analysis the coyote has apparently outwitted us all.

Another seemingly widespread plains animal that like the coyote somehow manages to remain almost undetected by most people is the spotted skunk (Figure 65). It is a smaller relative of the larger and much more generally seen striped skunk. It occurs throughout almost the entire region considered here, but is replaced by the very similar western spotted skunk in the Black Mesa country of extreme western Oklahoma. As its common name implies, its pelage pattern is one of elongated white spots and stripes on a black background. The bushy tail is mostly black, with contrasting white present especially on the underside, but the white portion becomes highly visible when the tail

Figure 65. Eastern spotted skunk, adult warning display

is erected during threat. Its Latin species name *putorius* gives some hint of the putrid smell this small skunk can discharge.

Like the striped skunk, the spotted skunk is essentially a woodland species, but in the eastern plains tends to occur around farms, feeding on a diversity of small vertebrates, especially mice, invertebrates, fruits, grains, and whatever else might be available. It is more nocturnal than the striped skunk and somehow seems to avoid being run over by traffic to the degree typical of striped skunks. Considerably more agile than the striped skunk, it is able to climb trees, and it is also more social than the striped skunk, with groups sometimes sharing winter dens. Adults don't maintain definable territories, but move about through home ranges, at least until females mature and establish a breeding den for giving birth and rearing their young.

Because of their effective defense system, spotted skunks have few enemies, of which the great horned owl is probably the major one. Horned owls will attack any skunk without hesitation, but spraying by a skunk may have serious effects, perhaps blinding the owl or at least rendering it almost unable to hunt again. If given fair warning, the spotted skunk will run toward an intruder, stop about ten feet away, then stamp its feet as a warning. If that fails, it will do a quick handstand and, still facing the enemy, twist its rump enough to eject a well-directed spray in its face. The musk contains an odorous sulfide, mercaptan, which is somewhat neutralized by ammonia solutions and other available household materials, such as those mentioned in the striped skunk account. Thus, skunks are best appreciated from a considerable distance. Further, at least striped skunks are a major carrier of rabies in the United States, giving additional reason for avoiding close encounters of a most unpleasant kind.

CHAPTER 11

The Transients

Migrants and Drifters

Years seem far too short for naturalists, and single seasons are even briefer. Springs come and go so rapidly that summer-blooming prairie roses seem to appear even before the blossoms of pasque flowers and purple avens have been transformed into wispy filaments of prairie smoke. Yet, for a few weeks each spring and fall, the Great Plains are visited from the air by uncountable millions of transients. Many move silently at night, with the only clear evidence of their passing the broken bodies of small nocturnal migrant birds lying at the bases of tall buildings. Perhaps, while watching the pinpricks of light from distant stars, they fail to see those colossal objects in front of them. Others we hear but don't see, as in the choruses of migrating waterfowl flying high above the reflected lights of our cities, intent on reaching some destination beyond our ken. Many are so large or move in such numbers that we cannot overlook them. We come to measure our springs and falls by their regular appearance. These comings and goings are often the natural guideposts of our lives, as in "the spring we saw the whooping cranes" or "the winter of the snowy owls." Such events seem to provide far more satisfying memory-points than, for example, "the year I turned forty."

Because the Great Plains offer an unobstructed flyway between the Gulf Coast and the arctic breeding grounds for many long-distance avian migrants, are without mountains to traverse, and have abundant foraging and resting opportunities along the way, migration here is almost without parallel, except perhaps along some northern coastlines. Beyond the regular seasonal migrants, there are those transients that at times appear unexpectedly and often disappear just as quickly. Such would include elusive creatures like mountain lions, wandering east out of Wyoming and Colorado, or the odd moose that sometimes shows up in cow pastures from North Dakota to as far south as Nebraska. Even more surprisingly, there are also occasional armadillos that have been seen ambling up the highways of northern Kansas and southern Nebraska. In spite of their seemingly impervious armor, these animals often end up as part of the highway's flattened fauna after encountering speeding trucks.

Among the most exciting of the Great Plains' rare to occasional guests are the large, northern-breeding raptors that periodically drift southward as their boreal forest and tundra food supplies run scarce. Such birds include the great gray owl, the gyrfalcon, and the snowy owl. For natives of North Dakota, a snowy owl hardly merits stopping one's car on a frigid winter day, but farther south these birds increasingly become subjects of frantic communication on state birding "hotlines." Likewise, the appearance of a true North American rarity, such as one of the few Eurasian "common" cranes that have mistakenly joined sandhill cranes (Figure 66) in Siberia and followed them across the Bering Sea and south to the Great Plains, can cause a nationwide alert among hard-core bird-watchers, bringing to Nebraska human migrants from coast to coast.

I have watched countless brant geese and eiders wedge their way northward along the edges of the "smoky sea," the fog-draped Bering Sea coastline, and have stood on the ice-bound edges of Hudson Bay, scanning the skies for flocks of snow geese returning to their vast nesting colonies. They are once-in-a-lifetime experiences that are never to be forgotten. But they don't match the Platte Valley in March.

Late March along the Platte River is sheer magic. For more than a month, sandhill cranes pour into the valley from Texas and New Mexico, their numbers topping out at nearly a half million. From then until about mid-April, countless cranes make their daily flights out into cornfields and meadows to harvest whatever grain and invertebrates they can find, building up critical fat stores that are essential for their remaining spring migration and arctic breeding. Near sunset each evening the birds begin moving toward traditional roosting sites on sandy islands and barren sandbars. Flocks of a thousand or more

Figure 66. Sandhill crane, adult

cranes fly low over the river, their voices rising and falling in crescendos as they approach, pass overhead, and disappear again in the distance. As the sun sinks ever closer toward the horizon, the birds become increasingly nervous, hoping to find a safe landing place before it is wholly dark. Finally, a single brave crane touches down, to be followed moments later by another, then dozens, and finally hundreds. Eventually as many as 20,000 may occupy a single roost, stretched out along a mile or more of the river. After some initial jostling for position and rejoining of any pairs or family members that may have been temporarily separated in the confusion of landing, darkness settles in on the crane roost.

In the sun-warmed mid-April days along the Platte, the birds ascend in great slow-motion whirlwinds, their wings lifted by invisible thermals until they are almost out of sight, with only their excited calls wafting down to betray their excitement as they head off toward unknowable destinations somewhere along the northern rim of the world. Soon after leaving the Platte Valley, the great flocks begin to split up, some heading for Hudson Bay shorelines and islands, others for the high-arctic tundras of far-northern Canada, still others to the Yukon-Kuskokwim delta of Alaska, and some even to Siberian tundras still three thousand miles away.

Although sandhill cranes once bred in the Dakotas and Nebraska, they were eliminated by the early 1900s, the last records being for 1916 in North Dakota, for 1910 in South Dakota, and for 1883 in Nebraska. However, in 1973, there was a North Dakota nesting, and during the late 1990s, nestings were documented in Nebraska as well as in Iowa. With luck and care, the unforgettable bugle-like calls of crane families may again become a regular part of our Great Plains breeding avifauna. Until then, we must bid them farewell during the ides of every April, as they head resolutely northward. After they have all gone, the Platte reverts back to being just an ordinary Great Plains river, waiting silently for another coming of the cranes.

Much larger and far rarer than the sandhill crane, the whooping crane (Figure 67) is the gold standard of American birds; it is at once the tallest, one of the rarest, and certainly one of the most beautiful of all North American bird species. I was already in my sixties before I finally saw a wild whooping crane in Nebraska. And yet, except for seeing them on their wintering grounds in coastal Texas, the Platte Valley of Nebraska has always represented the most likely place for seeing the

Figure 67. Whooping crane, adult

birds on migration. Of nearly six hundred Great Plains sighting records recently summarized by Jane Austin and Amy Richert (2001), the majority (60 percent) were for Nebraska, followed by 15 percent for Kansas, about 9 percent each for the Dakotas, and 5 percent for Oklahoma. There were also a few sightings for Montana and interior Texas.

In Kansas, Quivira National Wildlife Refuge and Cheyenne Bottoms State Wildlife Area sometimes attract whooping cranes, and in the Dakotas the birds mostly concentrate in and east of the Missouri River valley. In Nebraska it is the central Platte Valley that has historically been the whooping crane's favorite stopover point. I once calculated that about 80 percent of the historic Nebraska records for whooping cranes originally occurred in the immediate vicinity of the Platte River and within a stretch of about 75 to 100 miles extending west from Grand Island. In more recent decades, the birds have increasingly used the Rainwater Basin, an area of shallow wetlands south of the Platte, and some of the rivers and wet meadows in the Sandhills, north to the Niobrara River. Even during the years when whooping cranes were plentiful, their flock sizes were small; family-sized units are the norm for migrating whooping cranes.

The historic changes in whooping crane distribution in Nebraska probably are the result of wholesale and almost unregulated "dewatering" of the Platte during the twentieth century, greatly reducing its width and allowing woody vegetational growth to develop along its edges and sandbars. Whooping cranes like wider wetlands and less obstructed views than sandhill cranes will tolerate, and forage to a lesser degree on upland sites, so their reduced use of the Platte is understandable. These trends in river losses have at least been slowed down by federal mandates for minimum Platte flows to protect whooping cranes and other endangered or threatened species, but much of the damage that has been done to the Platte's historic ecology will be difficult to reverse.

Enormous federal support has been put into the recovery efforts for the whooping cranes, and over the past six decades these activities have slowly paid off. There is now a better chance of seeing whooping cranes in Nebraska than perhaps at any time during the twentieth century, and the population of wild birds has increased from fewer than 20 to nearly 200. Yet, the whooping crane is still not secure, and every year the main population must undertake a long and dangerous migration from breeding grounds in Canada's Wood Buffalo

Figure 68. Snow geese, adults

National Park to Aransas National Wildlife Refuge in coastal Texas. It is a journey fraught with danger, from accidental collisions with electric lines, attacks by coyotes or golden eagles, and death by sport hunters prone to shoot at anything larger than a breadbox. Currently, efforts are under way to establish some nonmigratory populations, such as one in southern Florida, where the hazards of migration would not exist. The success of these efforts is still uncertain.

Like the cranes, North American geese have been traversing the skies of the northern hemisphere on their seasonal migrations for as long as primates have been walking upright, and probably far longer. The snow goose (Figure 68) once had a wonderful scientific name, *Chen hyperborea*, that translates as "the goose from beyond the north wind." It is an image that has always seemed just perfect to me. I remember seeing what appeared to be impossibly high flocks of white geese during fall pheasant hunts with my father and uncles in the late 1930s. When I asked where they had come from or where they were going, I got only evasive answers. Indeed, the high Canadian nesting grounds of the snow goose, and especially its dark grayish blue variant, the so-called blue goose, were at that time barely understood. Writing in 1923, A. C. Bent stated, "No one knows where the blue goose goes to spend the summer, and none of the numerous Arctic explorers have ever found its breeding grounds." In 1932 George Sutton finally reported that Southampton Island, at the upper end of Hudson Bay, is a major nesting ground of the blue goose as well as of the typical white-plumage type of snow goose. After much additional study, it was learned that not only are the blue goose and the snow goose one and the same species, but also that they are rather simple genetic variants of one another.

Genetics aside, the sounds and sights of migrating snow geese have followed me throughout my life. Perhaps more accurately, I have followed them, for they have drawn me to both arctic Canada and arctic Alaska, as well as to the birds' wintering grounds from southern California to the Gulf Coast. They are the first of the geese to arrive back into the northern plains in early spring and the earliest to leave on their way farther north. They offer the first, uncertain evidence that the Dakota winter is nearly over. They constantly push their tolerance limits for cold, snow, and ice; snow geese they are called, and snow geese they are. I once tried to discover if there might be an Innuit legend that would provide an explanation as to why the birds are white. Eventually I found that the native people of the high arctic typically only offered folklore explanations for those animals that aren't white!

Besides geese and cranes, some of the longest migrations of large North American birds are those of a few hawks, especially the Swainson's hawk. This prairie-loving species might cover at least 10,000 miles in the course of a single year as it flies from the northern plains of Canada and the United States to the similar temperate grasslands of the Argentine pampas. Swainson's hawks occupy a somewhat precarious niche, situated both geographically and ecologically between the ferruginous hawk of the western high plains and badlands, and the riparian woods and upland hardwood forests of the American central lowlands and eastern states that are the domain of the red-tailed hawk. The Swainson's hawk needs both open, relatively short-grass plains for effective hunting, as does the ferruginous hawk, as well as moderate-sized trees for nesting, like the red-tailed hawk.

Swainson's are substantially smaller birds than either of these two congeners, and in matters of territorial disputes with either are likely to come out second-best. Their prey size also probably averages somewhat smaller than either, but estimates of breeding-season dietary overlap among the three indicate a very high incidence of species commonality. On the other hand, estimates of breeding-season habitat overlap are much lower. Furthermore, after the breeding season the Swainson's hawk rather rapidly shifts to a diet rich in grasshoppers and other large insects, whereas the other two species continue their diet of rodents and other small mammals. The usual abundance of grasshoppers in late-summer grasslands then makes such foraging easy and allows for a rapid postbreeding buildup of fat supplies.

Of these three species, only the Swainson's hawk is strongly migratory, and by late September they are grouping into flocks sometimes numbering in the hundreds and heading south. Swainson's hawks probably don't choose to gather in flocks for social reasons, but instead only select the same topography, visibility, and overall flying conditions, which bring thousands of them into common flight paths. For at least six weeks they funnel southward through Mexico, Central America, and northern South America, stopping little if at all to "refuel" and apparently living on their stored fat reserves. Eventually they reach the grasslands of southern South America, at the start of the southern spring, when again insect food supplies are becoming widespread. The absence of effective controls on pesticide use in Latin America has been disastrous for Swainson's hawks in recent years, and their populations have declined greatly. Swainson's hawks also suffer from the fact that they are much less wary than red-tailed hawks or ferruginous hawks. A perched Swainson's will often allow a car to coast up to as close as twenty to thirty feet before finally flushing. This behavior provides great photographic opportunities, but also often literally places the birds in the sights of people carrying rifles or shotguns.

Swainson's hawks return to the Great Plains in March and April, as warming skies generate thermal updrafts perfect for spring migration. Again, these birds are then more prone to be seen in flocks, called "kettles," than are any of the other grassland hawks. Some continue northward well beyond the Canadian border, to breed in the mixed-grass prairies of central Alberta and Saskatchewan. A very few breed as far north as northern British Columbia. Such birds have one-way migration distances of up to at least 6,000 miles. Probably only some arctic-breeding peregrines, which also regularly winter in temperate South America, would have significantly longer routes among North American raptors.

Arctic-breeding birds are often among the most sought-after species on the lists of avid bird-watchers; we either have to make the long, expensive trips to their arctic homes or hope that they might come into view during their annual passages to and from wintering areas. Such appealing birds that come to mind almost immediately include the snowy owl, gyrfalcon, Hudsonian godwit, Harris's sparrow, and whimbrel.

If the snow goose is named perfectly, then so too is the whimbrel. There is a hint of the arctic wind in its name, of its wild whistled calls, and of the wistful feeling that I always get when I see them, remembering the Canadian tundra in its full June glory. The name "whimbrel" is indeed thought to be imitative of its call, and its generic name *Numenius* refers to the shape of its slightly decurved bill, resembling that of a new moon.

I once set up a blind on a nesting whimbrel in the coastal tundra near Churchill, Manitoba. The bird was so tolerant of the blind and my comings and goings that I probably could have dispensed with it completely. I had my blind placed close enough to the nest that, through my telephoto lens, I could easily watch every mosquito that landed on its head and tried to find biting points at the base of its bill or among the short feathers immediately around its eyes. The bird suffered silently through this ordeal day after day, and I think I was as relieved as the female when it finally transformed its clutch of four eggs into four spindly legged babies. After drying off and resting for a few hours, their mother gathered them up and hustled them into the tundra, where they soon disappeared. Some arctic stories such as this one do end happily, but many more are terminated by arctic foxes, predatory gull-like birds called jaegers, other predators, or late blizzards. Many

Figure 69. Hudsonian godwit, adult

of the shorebird nests I was watching were covered one day by eight inches of fresh snow and had to be abandoned by the incubating adults, especially among the smallest sandpipers. Most larger birds, such as snow geese and whimbrels, were able to sit tight and outlast the snow. For the others it was too late to start over, and the long transequatorial flight southward was all that was left for them.

Hudsonian godwits (Figure 69) are rather smaller versions of the marbled godwit of the northern plains, and additionally have slightly more uptilted bills and a much more chestnut-rufous body, at least during the breeding season. Bird-lovers of the Great Plains are most likely to see them only during a rather brief period in late April and early May, as they push northward toward a relatively few known nesting grounds in northwestern and north-central Canada. One of their known nesting grounds is the Churchill area of northeastern Manitoba where, with whimbrels and a half dozen other species of sandpipers, they stake out territorial claims along the scraggly spruce tree line that marks the end of the boreal forest and the beginning of Canada's arctic tundra.

It usually comes as an unexpected shock for a person who has only seen Hudsonian godwits standing sedately about marshy shorelines or wading in the shallows of prairie wetlands to encounter these birds on their home turf. There, one is more likely to see a male perched precariously on the branch of a small conifer, screaming out a warning to its mate that a human intruder is approaching the nest. Because of this effective warning system, the chances of finding a godwit nest in the subarctic tundra are almost as slight as finding gold in the muddy and placid rivers of the Great Plains. Although in the course of about a week in the vicinity of Churchill, Manitoba, I stumbled across the nests of many wonderful shorebirds, the godwits always eluded me.

It is always something of a surprise when, in asking a friend from either coast about which birds they might most like to see during a visit to the Great Plains, the answer rather frequently given is the Harris's sparrow (Figure 70). Fulfilling such a wish is possible only from

Figure 70. Harris's sparrow, adult male

late fall to early spring, for these large and attractive sparrows breed along the forest-tundra tree line of central Canada. They visit the central portions of the Great Plains during the colder months but are mostly spring and fall migrants in the Dakotas. Where they overwinter they join with white-throated sparrows, American tree sparrows, juncos, and other migrant or resident sparrows, but largely avoid the backyard bird feeders of cities and suburbs. Instead they tend to gather around plum thickets and woodlot edges, foraging on weed seeds. Their plump size and pink bills help to separate them from house sparrows, to which they otherwise bear some plumage similarities. Another distinction is their voice; the male's "song" during winter is a plaintive, quavering whistle, seemingly with more than a touch of implicit sadness. Later it may utter multiple notes of this type, followed by a series on a higher or lower pitch. Every time I hear it I am immediately transported back to the Canadian arctic tree line.

In a manner something like the snow goose, the nesting grounds of the Harris's sparrow remained unknown until comparatively recently. When a railroad to Fort Churchill, on the west coast of Hudson Bay, was completed in the late 1920s, this rich and primeval area of tundra and forest edge was finally opened to ornithologists. In 1931 George Miksch Sutton, the man who discovered the Southampton Island nesting grounds of the blue-morph snow goose, spent late May and early June in search of breeding pairs, and finally in mid-June, after weeks of searching, he located an occupied nest. It was one of the last North American bird species to finally have its nesting grounds discovered. Forty years later, I likewise saw Harris's sparrows in the Churchill area, but was too preoccupied with watching shorebirds and waterfowl to try to seek out the sparrows' nests.

Snowy owls are birds that seem ready-made for attractive caricature—their faces lack the rather malevolent appearance of an angry great horned owl, and their plumage is so luxuriant and so fluffed out in cold weather as to make them look like avian snow creatures when they are standing on a snow-covered landscape. The adult males show only small flecks of gray on their otherwise immaculate white upperparts, whereas adult females and younger birds of both sexes are more heavily marked with grayish black. In both sexes, the enormous eyes glow like golden embers, surrounded by a circular disk of white. The black beak is almost hidden by a feathered mustache, and feathers also extend down to hide their toes and powerful talons. They are on average the largest and heaviest of our owls; females may weigh up to almost six pounds, which is considerably heavier than any great gray owl and more than nearly all great horned owls. With such power, the birds can kill and dismember young geese, adult ptarmigans, and even arctic hares. Although an adult arctic hare may weigh substantially more than even a female snowy owl (up to about twelve pounds), an owl won't hesitate to attack. According to one account, the owl will impale the hare's back, the long hind talon of one foot digging deep into its rib cavity and perhaps reaching its heart, while its other rear talon is anchored into the ground, holding the hare in place until it ceases struggling.

When snowy owls begin to run out of available prey on their tundra breeding grounds, the birds gradually ghost southward. They avoid the heavy coniferous forests of Canada, where hunting would be difficult for these open-country and visually hunting birds, and settle into the plains and prairies of southern Canada and the northern states. They rarely stray far south of the winter snow line, evidently aware of their conspicuousness on a snow-free landscape. The only snowy owl I have ever seen in Nebraska was one that spent most of an unusually snowy winter on the University of Nebraska campus in Lincoln, where it often perched at the top of our embarrassingly gigantic football stadium, getting a better panoramic view than that provided millionaires in their luxurious skyboxes. It seemed to me to be the best possible use for a university football stadium. There it could probably easily find cottontail rabbits on campus and various rodents along the nearby train right-of-way, which must have tasted just as good as lemmings and arctic hares. Quite possibly, the domestic cats that often prowled the same railroad tracks for mice also had shorter than expected life spans.

As a youngster I looked forward to late fall pheasant hunts in North Dakota, after the ground had become snow-covered, more in the hopes of seeing a snowy owl than flushing any pheasants. Their seasonal appearance told me that, like my beloved snow geese and like polar bears, arctic foxes, ptarmigans, and arctic hares, some animals may best survive in a place where it is safer to be the color of winter than of summer.

If snowy owls are the largest and most beautiful owls of the high arctic, then gyrfalcons occupy the comparable prize position among the high-latitude hawks. I saw only a single gyrfalcon as a youngster in North Dakota, and none again until I visited the arctic myself. It was one of those momentary views out the side window of our family car, as we were driving in the country on an early spring day. As always, I watched for hawks on telephone poles as they sped past my view; they were usually visible just long enough to determine if they were red-tailed hawks or rough-legged hawks. On this particular occasion, I saw a pale hawk on an approaching pole and assumed it would turn out to be one of the very pale red-tailed hawks commonly called Krider's hawks. Yet, in the moment we passed it, I saw the distinct facial mask that identified it as a falcon. It was too late to shout out to my father to stop the car, and moments later it was lost to view. It was like one of those moments later in adult life, when an incredibly beautiful woman is momentarily in view and then is lost in the crowd. All that remains is a flash of memory, to be replayed over and over again in future years, but always with the same frustrating result.

I next saw gyrfalcons in the Alaskan tundra of the Kuskokwim delta, where I was studying eiders. On a rocky outcrop about a mile from my tent, a pair of gyrfalcons was nesting. It was too far to easily walk, for that lowland area was more water than land, and in the necessary process of walking around ponds, the distance between any two points had to be multiplied two- or threefold. Each day I would longingly cast eyes over toward that promontory, but each day also realized that there were more important things to be done near camp. Eventually my days ran out, and another might-have-been experience was added to my list.

We now know that gyrfalcons slip down into the northern plains states perhaps every winter, especially moving into large grasslands such as the Sandhills of Nebraska, where jackrabbits provide just as suitable prey as do arctic hares farther north. But few birders venture out into the Sandhills during winter, and so actual occurrence records for gyrfalcons are frustratingly few. Unlike the snowy owls, the ghosts of my North Dakota winters past, I keep hoping that perhaps the gyrfalcon will become the ghost of a winter yet to come. All told, gyrfalcons seem to me to be more myth than reality, more supposition than substance. Yet, it would be boring if all goals were easily met, or all birds easy to see. Because they aren't, there is always the new place to explore, the new dream to engage in, and the new animal to be seen. What better natural world could one ask for?

As compared with such relative migratory giants as geese, cranes, owls, and hawks, the smallest migratory bird of the Great Plains, and one of the tiniest of all warm-blooded animals, is the ruby-throated hummingbird. Even with a heavy layer of fat, which is acquired prior to their spring and fall migrations, adult ruby-throats barely weigh four grams, so that at least five could be mailed for the price of a single first-class stamp, with a bit left over for the envelope. Not only are ruby-throats tiny, but they are also among the most beautiful of all birds, for their feathers, especially those of the male's throat (its "gorget"), are modified in a way that breaks up the available "white" light into its spectral components, absorbing most hues but reflecting back a laser-bright ruby-red. This intense concentration of color can be directed, by feather-spreading and body orientation, directly toward a female or another male. In the latter case, it functions as a male-to-male threat, intended to establish social dominance, and may precede an actual fight if the simple visual signal is inadequate. When directed toward a female, it may serve to identify the bird's sex and probably also acts as a courtship signal. It is impossible for humans to judge the intensity of this signal exactly, as many hummingbird gorgets reflect ultraviolet light as well as hues in the visible red-to-violet spectrum.

All of this plumage beauty is supplemented by aerial acrobatics that are dazzling in their quickness, including rapid power dives ending above the head of the other bird or rapid, horizontal movements at the other bird's eye level, producing a shuttle-like action like that of a ping-pong ball bouncing rapidly back and forth before one's eyes. For all of his effort, the male is unlikely to receive any obvious encouragement; the female is prone to fly away and try to avoid the male's further attentions. Most matings are rapid, seemingly spur-of-the-moment events, and once fertilized, the two birds are unlikely to encounter each other again. Instead the female goes off to construct her nest, a marvelous mixture of spider webbing, bits of tree bark, and often mosses or lichens. The entire structure is little bigger than a walnut and is so artfully concealed with bark and lichens as to blend perfectly with the branch on which it rests.

Perhaps four or five days are needed for such nest construction, which is then followed by the laying of two tiny white eggs no larger than peas, or about a half inch in maximum length. The female incubates these tiny eggs for sixteen days, dividing her time between keeping the eggs warm with her body heat and feeding frequently enough so that she doesn't die of starvation. Unlike nonbreeding hummingbirds, incubating and brooding females cannot become dormant at night, thus saving energy by reducing their body heat. Long, cold nights may bring the female critically close to draining her entire energy reserves, and at least in my experience hummingbird nests are often located where first light of the morning sun strikes the nest and female, warming it and her as rapidly as possible.

Most ruby-throats nest to the east and north of the Great Plains states, the birds passing northward in April and early May and southward in September and early October. Birds from all over eastern North America funnel south into the Gulf Coast area of eastern Texas, and from there south to Mexico and northern Central America. Many of these birds might travel at least 1,000 miles each spring and fall, their tiny brains often directing them back each spring to the very territory that they used the year before. In spite of their small size and fragility, the birds are too elusive for most predators to catch, and their annual survival rates are as high or higher than most songbirds. Not many hummingbirds have been banded and subsequently recovered, but band recoveries and recaptures have proven that some ruby-throats may live as long as nine or ten years.

CHAPTER 12

What Is Still So Great about the Great Plains?

Although some parts of the natural world, such as the Sahara Desert, are ever increasing in size as a result of global warming and land misuse, the historic grassland ecosystems of the Great Plains are shrinking. At one time, the grasslands of the Great Plains may have blanketed about a million square miles of interior North America, extending from what is now southern Alberta in the northwest to coastal Texas in the south, and eastward in the upper Mississippi Valley to Lake Michigan and somewhat beyond, where oak savannas and hardwood forests progressively competed with and eventually supplanted the moist tallgrass prairies. The tallgrass prairies, which once probably covered at least 230,000 square miles, have since been reduced to perhaps 2 to 4 percent of their original area, with about 8,000 to 10,000 square miles now left between North Dakota and Texas, plus additional remnants in southern Manitoba and some tiny scattered fragments in the upper Mississippi Valley.

The Flint Hills of Kansas, at the western boundary of the tallgrass prairies, are easily the largest remaining area of intact tall grasslands south of Canada, with perhaps 5,000 square miles still fairly undamaged. There, more than 90 percent of the region is devoted to cattle production, because traditional farming is impossible on this rocky landscape. About 80 species of mammals, 300 species of birds, and 700 species of plants have been reported from the collective North American tallgrass prairies, although the number of characteristic organisms found at any one location is far fewer. In the tallgrass prairies of south-eastern Nebraska, for example, 350 to 400 species of plants may be found in larger stands of high-quality tallgrass, and most remnants of moderate size have over 200 species. In Kansas's Konza Prairie, with a similar vegetational diversity, over 30 species of reptiles and amphibians, 40 native mammals, and more than 200 bird species have been recorded within its roughly thirteen-square-mile area. Of its birds, about sixty are breeders and thirty of these are clearly associated with grasslands.

The mixed-grass prairies have been reduced to something like 35 to 45 percent of their original size, which was once in excess of 200,000 square miles and considerably more if the transitional short-grass–mixed-grass region is included. The present area of mixed-grass prairies from the Canadian border south through Nebraska consists of about 26,000 square miles, and at least as much additional acreage occurs from Kansas to Texas. The Canadian fescue grasslands of Alberta and Saskatchewan account for the rest of the surviving North American mixed-grass prairies, and these are still largely intact. Good-quality stands of mixed-grass prairie in the loess hills of south-central Nebraska sometimes contain nearly 250 species of plants, and the loess hills of western Iowa support at least 44 species of mammals and 36 species of reptiles and amphibians. About 100 species of birds are believed to nest in this general area, and 30 are common breeders.

The large (about 20,000 square miles) but much more fragile Nebraska Sandhills grasslands, although not true mixed-grass prairie, support many of the same species. Over 600 plant species occur in the Sandhills collectively, and good stands of dry valley or upland Sandhills prairies may have nearly 100 plant species. Higher floral species diversity occurs in the moist fens, which typically support 110 to 135 species. The Sandhills region collectively also supports about 30 species of reptiles and amphibians and 55 mammals. There are also nearly 150 breeding birds in this region, although most of the breeding birds are actually wetland- or woodland-dependent species, and only about 30 are true grassland associates. Single stands of upland Sandhills prairie usually support 6 to 9 breeding bird species and a similar number of small mammals, mainly arid-adapted and seed-eating rodents.

Breeding birds of the tallgrass and mixed-grass prairies that have declined the most rapidly in recent decades are (in diminishing rates of severity) Henslow's sparrow, Sprague's pipit, grasshopper sparrow, short-eared owl, lark sparrow, eastern meadowlark, Baird's sparrow, chestnut-collared longspur, bobolink, and dickcissel. Short-grass prairie and shrubsteppe birds that have correspondingly declined at the greatest rates in recent decades are the loggerhead shrike, Brewer's sparrow, Cassin's sparrow, horned lark, scaled quail, lark bunting, long-billed curlew, and western meadowlark.

Of the major prairie types, only the arid short-grass prairies of the western Great Plains retain a few locations that still provide some boundless sense of how they once might have appeared when bison and pronghorns roamed them in the tens of millions. Because the short-grass–mixed-grass boundary is a dynamic one that shifts with changing climates, the historic size of the short-grass prairies is especially difficult to judge. One estimate of about 240,000 square miles has been made on the basis of a historical map prepared by A. W. Küchler (1966). A much more restrictive estimate of its original primary area has been provided by the size of the Short-Grass Prairie Bird Conservation Region. This region consists of 148,000 square miles within eight states, reaching from southwestern South Dakota and southeastern Wyoming south to eastern New Mexico and western Texas (approximately the crosshatched area of Figure 1). It is one of thirty-six North American bird conservation regions that have been established to facilitate conservation efforts via the North American Bird Conservation Initiative, a consortium of governmental agencies and private organizations.

Based on recent estimates of Scott Gillihan and Michael Carter (2001), roughly half of the Short-Grass Prairie Bird Conservation Region still exists as recognizable grassland, but about 70 percent of it is privately owned, and only a small minority of the remainder (20 percent in Colorado) is not seriously degraded. Accepting and extrapolating these very rough estimates, it may be that within this region about 10,000 square miles of short-grass prairie still exists with a substantial degree of its original biotic composition, and perhaps another 50,000 to 60,000 square miles occur in a variably degraded state. Added to this would be the short-grass–mixed-grass prairie mosaics of Saskatchewan-Alberta and eastern Montana, and the shrubsteppes of northeastern and east-central Wyoming, these peripheral short-grass areas perhaps totaling another 100,000 square miles.

Of all the grassland wildlife, some of the classic short-grass mammals (black-footed ferret, swift fox, black-tailed prairie dog) have declined most precipitously. Given the seemingly large areas of short-grass prairies still remaining, this trend seems hard to explain. Although it does not need large areas for survival, the black-tailed prairie dog has been ruthlessly eliminated from nearly all of its onetime range, namely from an original Great Plains distribution of 150,000 to 400,000 square miles to a fraction of 1 percent of that area. Land conversion, sylvatic plague, and federal- and state-financed control programs have all contributed to the wholesale destruction of prairie dogs. Their ecological counterparts such as the black-footed ferret and swift fox have suffered associated downward population trends. The ferret is already listed as federally endangered, and the prairie dog and swift fox are currently candidates for federal listing as threatened species.

The larger short-grass mammals have suffered at least as much as the smaller ones. Uncontrolled hunting of the pronghorn, elk, bison, gray wolf, grizzly bear, and the now extinct badlands and high plains race of the bighorn eliminated them from the Great Plains long before grassland fragmentation and degradation became a significant factor in their demise. The elk, bison, bear, wolf, and bighorn are now either gone from the plains or are largely confined to small, protected areas, and the pronghorn has mostly retreated to arid sagebrush steppes.

The mountain plover, ferruginous hawk, burrowing owl, and Mc-Cown's longspur were also once part and parcel of this close-knit prairie dog–short-grass steppe assemblage. Their populations are probably also still declining, but most of these species have already become so rare that statistically significant population trends are hard to establish. In the last few years, the ferruginous hawk actually seems to have recovered slightly. Besides these four, eleven additional short-grass birds are classified as priority species of conservation concern, as determined in a database developed by a governmental-private consortium of ornithologists and conservationists called Partners in Flight.

One irony of the short-grass ecosystem is that the best remaining elements of it, the national grasslands, are largely the product of failed farming and ranching efforts during the first half of the twentieth century. The seven national grasslands in the five-state Rocky Mountain region of the U.S. Forest Service (which extends from Montana and North Dakota south through Colorado and Nebraska) total 3,750 square miles, and those located in Kansas, Oklahoma, and Texas add another 470 square miles. Collectively, over 4,000 square miles of short-grass prairies are now under Forest Service control and management. Thus, probably about 1.5 to 3.0 percent of the original roughly 150,000 to 250,000 square miles of typical Great Plains short-grass prairies are now encompassed within the national grasslands. Additional Bureau of Land Management lands in eastern Montana and Wyoming occur within the historic short-grass prairie region. All told, the federal government administers more than 32,000 square miles of arid grasslands and shrubsteppes in the American West, including the sagebrush steppes. Many of these federally managed lands have been degraded by pressures from ranchers who want even greater and still cheaper grazing rights, and who would prefer to see fewer prairie dogs (and their associated wildlife) but far more cattle or sheep on our publicly owned lands.

Where can one still go to see the kinds of Great Plains wildlife I have described in this book's pages? Appendix 2 provides a list of nearly 100 nature preserves and natural areas that are located within the six-state limits of this book, including private, state, and federal sites. Each of the states covered in this book has as well one or more available guides to locations for general wildlife watching, also cited in Appendix 2. A listing of more than 100 grassland preserves that are located within the entire native prairie region of southern Canada and the central and midwestern states and that might offer some wildlife-watching opportunities was also published in my earlier *Prairie Birds: Fragile Splendor in the Great Plains.*

National wildlife refuges, national parks, national grasslands, and nature sanctuaries of moderate to large size are especially good for locating and observing grassland and wetland wildlife. One of the most useful sources for locating birds in such locations is John O. Jones, *Where the Birds Are: A Guide to All 50 States and Canada.* It includes seasonal species status information derived from 210 refuge and park bird lists as well as less detailed information on more than 2,000 other birding locations. Sketch maps and narrative site information are often also provided.

With some effort and by visiting such localities, wonderful prairie landscape views can still be found within the Great Plains. The sounds of coyotes, elk, and bison can likewise be heard, the scents of fragrant wildflowers may be savored, and head-high grasses can be lightly caressed. There are still a few places where one can find and sit on a giant glacial-worn and lichen-capped boulder, with nothing but tall prairies visible to the far horizon, and imagine that only about 20,000 years ago mammoths might have rubbed their hairy sides against this very rock. Such marvelous moments are not easily gained. But then, nothing wonderful should be cheaply attained; were it so, the prize would seem far less rewarding and the world much less interesting. Each season in the Great Plains brings new potential delights, from snowy owls and gyrfalcons suddenly materializing in the dead of winter to uncountable bird migrations flooding the skies each spring and fall, with a riot of summer wildflowers and prairie grasses sandwiched in between as an extra bonus.

Armed with an inquiring mind and a field guide or two, anybody can experience these wonders and thereby expand his or her knowledge of the natural world by simply venturing out into the countryside and quietly becoming a part of it. Even if the animals themselves cannot be seen or heard, and this is often the case with small, nocturnal mammals, trying to identify them and their activities by their footprints left in snow or mud produces challenging mind-games. Similarly, single dropped feathers are sometimes identifiable as to species, and examining the regurgitated pellets left at owl and hawk roosts and nests can provide hours of pleasure in trying to understand the silent stories they can often tell. Only a rich imagination and an inquiring mind are required.

Let a man decide upon his favorite animal, and make a study of it—
let him learn to understand its sounds and motions.
Brave Buffalo (Teton Sioux), "Teton Sioux Music,"
Bureau of American Ethnology Bulletin 61 (1918).

Footprints, Hoofprints, Rump Patterns, and Antlers of Great Plains Species

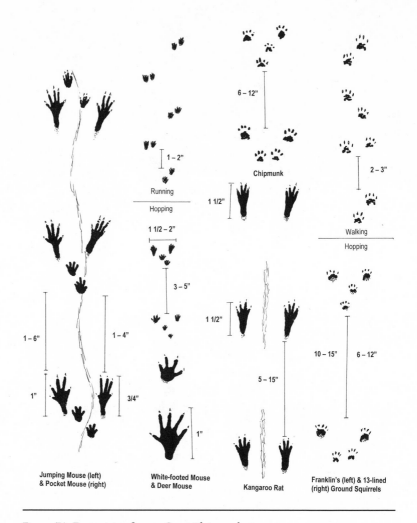

Figure 71. Footprints of some Great Plains rodents

214

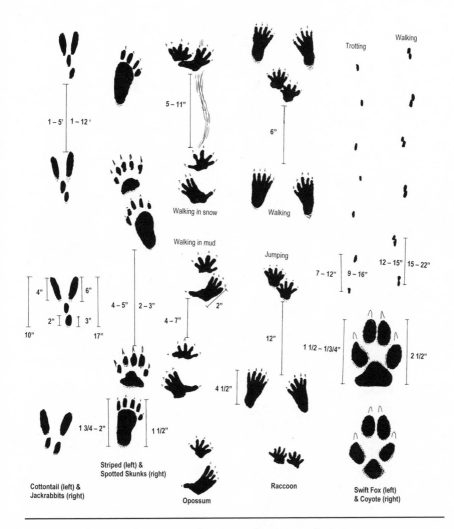

Figure 72. Footprints of some nonrodent Great Plains mammals

215

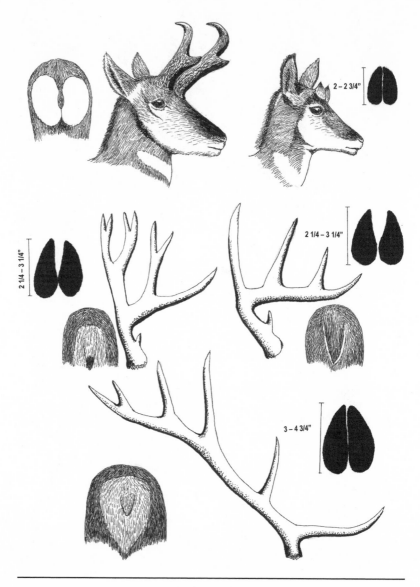

Figure 73. Pronghorn, heads of adult (left) and immature (right) males, plus species' rump pattern and hoofprint. Hoofprints, rump patterns, and antlers of male mule deer (middle left), white-tailed deer (middle right), and elk (bottom) are also shown for comparison.

APPENDIX *2*

Nature Preserves and Natural Areas in the Great Plains States

The following list of preserves, refuges, and natural areas includes all the larger federally owned wildlife refuges, monuments, parks, and grasslands in the Great Plains states from North Dakota south through Oklahoma and the Texas panhandle. It also includes many state, local, and private preserves having substantial natural habitats, especially sites larger than 1,000 acres. Areas (in acres) are indicated. This list overlaps with one of native grassland preserves published earlier (Johnsgard 2001a), but is more geographically restricted and includes woodland or wetland habitats as well as general nature sanctuaries.

Most state highway maps indicate exact locations for the federal preserves and state-owned sites, such as national wildlife refuges (N.W.R.) and state parks (S.P.), but city-, county-, and privately owned sites are often harder to locate. Phone numbers and/or addresses are provided whenever possible, but telephone area codes are subject to change and may need verification. County locations are indicated for those sites that are difficult to find on most maps. Locality information and complete bird lists for many of the listed sites may also be found in Jones (1990). Federal wetland management districts (W.M.D.), wildlife management areas (W.M.A.), and state recreation areas (S.R.A.) can often be located by contacting the appropriate federal office or state conservation agency (see below). Various sites may also be closed during certain seasons, and admission fees are charged at others (as noted). Some of the Nature Conservancy's (T.N.C.) preserves require permission for entry.

State natural heritage programs document the status of their floras and faunas, especially of rare species, and are a useful source of information. Those for the Dakotas and Nebraska are located within their respective state game departments. The Kansas program is part of the Kansas Biological Survey,

217

Raymond Nichols Hall, University of Kansas, Lawrence, KS 66045 (ph. 913/864-3453), and Oklahoma's is part of the Oklahoma Biological Survey, Sutton Hall, University of Oklahoma, Norman, OK 73019 (ph. 405/325-1985).

Those sites with available bird checklists are identified in the following list, giving species totals whenever possible. Mammal and herptile lists are also indicated when their existence is known. Biological checklists for many of these same sites are also available via the Northern Prairie Wildlife Research Center's highly informative website: *http://www.npwrc.usgs.gov/resource/othr data/chekbird/chekbird.htm.*

The U.S. Fish and Wildlife Service (USFWS) website is *http://www.fws.gov/,* and that of its regional field offices is *http://fws.gov/where/regfield/.html.* This site has hundreds of links to websites for individual refuges, wetland management districts, and so on, as well as to those of individual state agencies. Information on federally managed sites from North Dakota through Kansas (USFWS Region 6) is also available from the Bureau of Sport Fisheries and Wildlife, Box 25486, Denver Federal Center, Denver, CO 80225 (ph. 303/236-7398).

Other websites of special value to persons who may want further information are those of the National Park Service, *http://www.nps.gov/,* and the U.S. Forest Service, *http://www.fs.fed.us/.* The National Wildlife Federation's website, *http://www.nwf.org,* provides information on several endangered grassland species. The Nature Conservancy has a national website, *http:///www.tnc.org,* as well as individual state sites. The International Association of Fish and Wildlife Agencies' website provides links to the websites of individual state conservation agencies, *http://www.sso.org/iafwa.* The National Audubon Society also has a website with links to individual state chapters and their associated sanctuaries, *http://www.audubon.org.*

North Dakota (see also Krue 1992)

1. Arrowwood N.W.R. 15,934 ac. Ph. 701/285-3341. Bird list available (266 spp., 124 nesters). R.R. 1, Pingree, ND 58476. Upland mixed-grass prairies with several small lakes and a marshy impoundment of the James River. Nearby Chase Lake N.W.R. (375 ac.) is administered here.

2. Audubon N.W.R. 14,735 ac. Ph. 701/442-5474. Bird list available (205 spp., 85 nesters). R.R. 1, Coleharbor, ND 58531. Upland prairies and marshes adjoining a large impoundment.

3. Cedar River National Grassland. 6,700 ac. Ph. 605/374-3592. P.O. Box 390, Lemmon, SD 57638. Short-grass and mixed-grass prairie.

4. Crosby W.M.D. 85,819 ac. Ph. 701/965-6488. Box 148, Crosby, ND 58730. Prairie wetlands, including 92 waterfowl production areas, plus Lake Zahl. See also Devils Lake, Kulm, and Valley City W.M.D.

5. Cross Ranch Nature Preserve (T.N.C.). 6,000 ac. Ph. 701/794-8741. Near Bismarck. Mixed-grass upland prairie and floodplain. Unpublished bird list

available (see website above). Box 112, Hensler, ND 58547. Other T.N.C. holdings include John E. Williams Memorial Preserve, 1,601 ac. of mixed-grass prairie near Turtle Lake; Pigeon Point Preserve, 600 ac. of forest wetlands in Ransom Co.; and Sheridan Preserve, 1,437 ac. of prairie wetlands in Sheridan Co. For information on T.N.C. holdings in North Dakota, contact their field office, 2000 Schafer St., #8, Bismarck, ND 58501 (ph. 701/222-8464).

6. Des Lacs N.W.R. 18,881 ac. Ph. 701/965-6488. Bird list available (293 spp., 150 nesters). Box 578, Kenmare, ND 58746. Prairie (about 7,000–10,000 ac.), brushland, marshes, and several small impoundments.

7. Devils Lake W.M.D. 197,555 ac. Ph. 701/965-6488. 218 W. 4th St., Devil's Lake, ND 58301. Wetlands managed as waterfowl production areas. Includes Lake Alice N.W.R., Sully's Hill National Game Preserve, and Kelly Slough N.W.R. Nearby is the University of North Dakota's Oakville Prairie, near Emerado, with 40 spp. of breeding birds, 21 mammals, and 4 herptiles. West of Grand Forks is Turtle River S.P., 748 ac. (ph. 701/594-4445), located along a shoreline remnant of glacial Lake Agassiz. Relict prairie and bottomland hardwood forest.

8. J. Clark Salyer N.W.R. 58,693 ac. Ph. 701/768-2548. Bird and mammal lists (combined Souris Loop refuges) available (266 bird spp., 147 nesters). Box 66, Upham, ND 58789. Mixed-grass prairie, wet meadows, woods, and wetlands.

9. Kulm W.M.D. 42,352 ac. Ph. 701/647-2866. Box E, Kulm, ND 58456. Wetlands managed as waterfowl production areas.

10. Lake Alice N.W.R. 10,772 ac. Ph. 701/662-8611. Box 908, Devil's Lake, ND 58301. Lake surrounded by prairie and shoreline hardwoods.

11. Lake Ilo N.W.R. 3,737 ac. Ph. 701/548-4467. Bird list available (224 spp., 83 nesters); also 38 mammals, 6 reptiles, and 4 amphibians. Near Dunn Center, in Dunn Co. Managed by Des Lacs N.W.R., Kenmare, ND 58746. Lake surrounded by short-grass prairie and shoreline hardwoods.

12. Little Missouri National Grassland. 1,027,852 ac. Ph. 701/225-5152. Rt. 3, Box 131-B, Dickinson, ND 58601. Short-grass plains and badlands; the largest area of federally protected short-grass plains in the nation.

13. Long Lake N.W.R. 22,300 ac. Ph. 701/387-4397. Bird list available (206 spp., 78 nesters). R.R. 1, Moffit, ND 58560. Upland prairie, marshes, and cultivated lands surrounding Long Lake.

14. Lostwood N.W.R. 26,747 ac. Ph. 701/848-2722. Bird list (combined Souris Loop refuges) available (266 spp., 147 nesters); also lists of 37 mammals, 3 reptiles, and 4 amphibians. R.R. 2, Box 98, Kenmore, ND 58746. Upland prairie, aspen groveland, and alkaline lakes and marshes. Sprague's pipits, Baird's sparrows, and other mixed-grass prairie species are common.

15. Sheyenne National Grassland. 70,180 ac. Ph. 701/683-4342. Sheyenne Ranger District, P.O. Box 946, 701 Main St., Lisbon, ND 58054. Sandhills tallgrass prairie and riverine hardwoods on an ancient glacial delta of the Sheyenne

River; the largest area of federally owned tallgrass prairie in the United States. A remnant flock of greater prairie-chickens occurs here, along with sharp-tailed grouse. Nearby are the Mirror Pool North and Mirror Pool Swamp natural areas (471 ac.), with wetlands, deciduous forest, and dunes, and Pigeon Point Preserve (560 ac.), with spring-fed wetlands. Contact the Nature Conservancy, 2000 Schafer St., Suite B, Bismarck, ND 58501 (701/222-8464). Chapman, Ziegenhagen, and Fischer (1998) describe these Sheyenne delta prairies.

16. Sully's Hill National Game Preserve. 1,674 ac. Ph. 501/766-4272. Unpublished bird list of over 200 spp. Managed bison and elk herds, nature trail, and education center. Lake Metigoshe S.P. is nearby, and the Turtle Mountain region supports nesting buffleheads and moose. Rt. 1, Box 359, Bottineau, ND 58318.

17. Tewaukon N.W.R. 8,444 ac. Ph. 701/724-3598. Bird and mammal list available (236 bird spp., 98 nesters). R.R. 1, Cayuga, ND 58013. Prairie, cropland, marshes, and riverine wetlands on glacial till.

18. Theodore Roosevelt National Park. 70,416 ac. Ph. 701/623-4466. Bird and mammal lists available (170 bird spp., 58 nesters). P.O. Box 7, Medora, ND 58645. Eroded badlands, short-grass plains, and shrubsteppe. South of the South Unit of the park is Sully Creek S.R.A. (80 ac.), a small area providing canoe access to the Little Missouri River. Near the North Unit is Little Missouri S.P., an area of 5,749 ac. mostly accessible only by foot or horseback (ph. 701/328-5357).

19. Upper Souris N.W.R. 32,000 ac. Ph. 701/845-3466. Bird and mammal lists (combined Souris Loop refuges) available (266 bird spp., 147 nesters). R.R. 1, Foxholme, ND 58738. Mixed-grass prairie, woods, and wetlands.

20. Valley City W.M.D. 16,198 ac. Ph. 701/845-3466. R.R. 1, Valley City, ND 58072. Wetlands and grassland managed for waterfowl, in 72 units. Hundreds of waterfowl production areas (totaling almost 250,000 ac.) are located in this rich "prairie potholes" region of central North Dakota. North Dakota also has about 190,000 ac. of state-managed wildlife management areas, mostly open to public use. For information, see the Northern Prairie Wildlife Research Center's website shown above.

South Dakota (see also Pettingill 1981, Peterson 1993)

1. Badlands National Grassland. 243,302 ac. Ph. 605/433-5361. Bird list available (208 spp.). P.O. Box 6, Interior, SD 57750. Eroded high plains with short-grass and mixed-grass prairie and shrubsteppe. Interpretive center at Wall.

2. Black Hills National Forest, 1,235,453 ac., and Bear Butte State Park. Ponderosa forest and high plains. Bird list of 188 spp. (117 nesters) for Bear Butte S.P. The total Black Hills list includes at least 226 spp. (Pettingill and Whitney 1965), and a recent bird-finding guide exists (Peterson 1993). Contact National Forest office, P.O. Box 792, Custer, SD 57730 (ph. 605/673-2251), or the Bear Butte park office (ph. 605/347-5240).

3. Buffalo Gap National Grassland. 591,771 ac. Ph. 605/279-2125 or 605/745-4107. Bird list available (197 spp.). P.O. Box 425, U.S. Forest Service, 708 Main St., Wall, SD 57790, or U.S. Forest Service, 209 N. River, Hot Springs, SD 57747. Short-grass and mixed-grass prairie. Supports swift foxes, black-footed ferrets (reintroduced), prairie dogs, pronghorns, and other typical short-grass species. For contiguous Oglala National Grassland, see Nebraska listing.

4. Crystal Springs Prairie Preserve (T.N.C.). 1,920 ac. Ph. 701/222-8464. Near Clear Lake, Duel Co. Tallgrass prairie.

5. Fort Pierre National Grassland. 115,996 ac. Ph. 605/224-5517. P.O. Box 417, 124 S. Euclid Ave., Pierre, SD 57501. Short-grass and mixed-grass prairie. Supports greater prairie-chickens and sharp-tailed grouse.

6. Grand River National Grassland. 156,000 ac. Ph. 605/374-3592. P.O. Box 390, Lemmon, SD 57638. Short-grass and mixed-grass prairie.

7. Lacreek N.W.R. 16,250 ac. Ph. 605/685-6508. Bird and mammal lists available (273 bird spp., 93 nesters). Star Rt. 3, Martin, SD 57551. Sandhills grasslands, including about 4,000 ac. of mixed-grass prairie and wetlands in the Little White River valley. Trumpeter swans breed here.

8. Lake Andes N.W.R. 5,942 ac. Ph. 605/487-7603. Bird list available (214 spp., 85 nesters). R.R. 1, Box 77, Lake Andes, SD 57356. Tallgrass prairie, woods, and wetlands.

9. Madison W.M.D. 25,000 ac. Ph. 605/256-2974. Bird list of 297 spp. Box 48, Madison, SD 57042. Prairie wetlands managed for waterfowl over a nine-county area.

10. Pocasse N.W.R. 2,585 ac. Ph. 605/885-6320. Along upper part of Oahe Reservoir, in Campbell Co. Marshy shorelines and upland grasslands; an important stopover point for migratory waterfowl.

11. Samuel H. Ordway Jr. Memorial Prairie Preserve (T.N.C.). 7,800 ac. Ph. 605/439-3475 or 701/222-8464. Near Leola, Brown Co. Tallgrass prairies and wetlands; the largest tallgrass prairie in South Dakota. Other state T.N.C. preserves include Crystal Springs in Deuel Co. (see above); Hansen Nature Preserve, 800 ac. in Brown Co.; Sioux Prairie, 200 ac. in Moody County; E. M. and Ida Young Preserve, 901 ac. in Sully Co.; and Wilson Savanna, 160 ac. of oak savanna in Lincoln Co. For information on South Dakota T.N.C. holdings, contact their field office, 1000 West Ave. North, Sioux Falls, SD 57104 (ph. 605/331-0619). State parks include Union County S.P., 499 ac., with eastern species such as whip-poor-will and scarlet tanager. Newton Hills S.P., 948 ac., along the Big Sioux River, Lincoln Co., has these same species and American woodcock. The Adams Nature Preserve, 1,500 ac., in the extreme southeastern Missouri Valley near Vermillion, has grasslands, wooded bottomlands, and seven miles of trails. Beaver Creek Natural Area (165 ac.) is nearby (ph. 605/773-3391). Contact Division of Parks and Recreation, 523 E. Capitol Ave., Pierre, SD 57501.

12. Sand Lake N.W.R. 21,451 ac. Ph. 605/885-6320. Bird list available (263 spp., 111 nesters); mammal and herp lists are also available. R.R. 1, Columbia, SD 57433. Up to 2 million snow geese may occur during migration, and Franklin's gulls nest in vast numbers. Tallgrass prairie and woods bordering the James River; two lakes and several marshes; the total water area is about 11,000 ac.

13. Waubay N.W.R. 4,650 ac. Ph. 605/947-4521. Bird list available (244 spp., 109 nesters). Box 79, Waubay, SD 57273. Tallgrass prairie and dozens of glacial-formed wetland potholes on rolling upland till. Many shorebird, grebe, and duck species typical of the Dakota "prairie pothole" region nest here. Nearby is Sica Hollow S.P. (807 ac.), Marshall Co., excellent for aquatic and eastern hardwood birds, including the pileated woodpecker. Big Stone Island Natural Area (100 ac.) is also nearby at the south end of Big Stone Lake; this lake and adjoining woods are a major attraction to migrating birds. This glaciated region supports hundreds of wetlands designated as waterfowl production areas (WPAs); the largest numbers are in Day Co. (34), followed by McPherson Co. (25), Lake Co. (24), and Roberts Co. (21). In spring these attract many waterfowl and shorebirds.

14. Wind Cave National Park, 28,056 ac., and adjacent Custer S.P. (about 73,000 ac.). Bird list, including Black Hills National Forest, available (186 spp.). Short-grass prairie and coniferous forest. Contact U.S. Forest Service, Hot Springs, SD 57747 (ph. 605/745-4600). Wind Cave N.P. and Custer S.P. (ph. 605/255-4515) are partly managed for bison and other large ungulates. See also Black Hills National Forest account above.

Nebraska (see also Johnsgard 2001b, Krue 1997)

1. Agate Fossil Beds National Monument. 2,700 ac. Ph. 308/436-4340 or 308/668-2211. P.O. Box 427, Gering, NE 69341. Short-grass plains with Cenozoic fossil deposits from about 20 million years ago. Interpretive center.

2. Boyer Chute N.W.R. 2,500 ac. Ph. 712/642-4121. A new refuge located along the Missouri River, separated from the shore by a narrow channel ("chute"). About 5 miles of wooded trails. Located east of Fort Calhoun and managed by DeSoto N.W.R., a 7,800-ac. refuge 5 miles northwest. On the Iowa side of the Missouri River, DeSoto N.W.R. is primarily managed for snow geese in the fall; up to 800,000 birds may be present in late October and November. Bald eagles are then often abundant. An interpretive center is open year-round.

3. Crescent Lake N.W.R. 45,818 ac. Ph. 308/762-4893. Bird and mammal lists available (279 bird spp., 86 nesters). Star Rt., Ellsworth, NE 69340. Sandhills grasslands and dozens of small, mostly alkaline lakes and marshes. These wetlands support many nearly unique nesters, including cinnamon teal, American avocet, black-necked stilt, and white-faced ibis.

4. Fontenelle Forest Nature Center. 1,311 ac. Ph. 402/731-3140. 1111 N. Bellevue Blvd., Bellevue, NE 68005. A near-virgin hardwood forest along the Missouri River. Lists of birds (over 250 spp.) and mammals available. One of the best riverine hardwood forests surviving in the state, supporting several rare breeding bird species such as the parula, yellow-throated and prothonotary warblers, and the only known nesting pileated woodpeckers in Nebraska. Seventeen miles of trails, a new educational Learning Center, and an observation blind.

5. Fort Niobrara N.W.R. 19,124 ac. Ph. 402/376-3789. Bird and mammal lists available (225 bird spp., 76 nesters). Hidden Timber Rt., Valentine, NE 69201. Riverine woods, grassy uplands, and sandhills prairie, managed for bison and elk. Burrowing owls and other grassland birds such as sharp-tailed grouse are common, and the Niobrara River flows through the refuge.

6. Indian Cave S.P. 3,000 ac. Ph. 402/883-2575. A highly wooded stretch of the Missouri River in extreme southeastern Nebraska, supporting many species with southern affinities, such as whip-poor-wills, chuck-wills-widows, and flying squirrels. There are 20 miles of hiking trails, some of which are very steep. Timber rattlesnakes and copperheads both occur here. Not far south, on the Missouri side of the river near Mound City, is Squaw Creek N.W.R., about 7,000 ac. Ph. 660/442-3187. Refuge checklist of 263 bird spp., including 104 breeders. Large migrations of waterfowl gather here each spring and fall.

7. Lake McConaughy. This large reservoir (about 36,000 ac.) near Ogallala attracts vast numbers of migrating birds in spring and fall, with many waterfowl and bald eagles overwintering. The local bird list of more than 340 spp. is the largest in the entire Great Plains north of the Texas coast, and one of the largest in the nation (Brown and Brown 2001). The lake is bounded by the Nebraska Sandhills on the north and by cedar-lined canyons on the south, where the University of Nebraska's Cedar Point Biological Station is located. Below the dam in the river floodplain are typical riverine hardwood forests.

8. Lillian Annette Rowe Sanctuary (National Audubon Society). 1,600 ac. Ph. 308/468-5282. 44450 Elm Island Rd., Gibbon, NE 68840; located on Elm Island Road, 2 miles south and 2 miles west of I-80 exit 285. An interpretive center should open in 2003, and there is a summer hiking trail. Blinds are available for guided crane-viewing from early March to early April (reservations needed; $15 fee; email: *rowe@nctc.net*). Nearby public-access crane-viewing locations include a viewing platform located beside a bridge over the widest channel of the Platte, 1 3/4 miles south of exit 285, and on the Hike-Bike Trail bridge over the Platte at the Fort Kearny S.R.A. campground, 2 miles south and 4 miles east of exit 272. See also Platte River crane-viewing sites listed below. Lists of Platte Valley birds, mammals, and herps were provided by Johnsgard (1984) and are also available on the Northern Prairie Research Station website mentioned at the beginning of this list.

9. Neale Woods Nature Center. 554 ac. Ph. 402/453-5615. Owned by the Fontenelle Forest Association; located at the north end of Omaha, on Edith Marie Ave. A bird list of 190 spp. is available. Forested floodplain and hills, restored prairie, and an interpretive center. Nine miles of hiking trails. Admission fee.

10. Nebraska National Forest and Pine Ridge National Recreation Area. Three rather widely separated units totaling 360,000 ac. and including the Bessey District, 90,448 ac., the Samuel McKelvie District, 115,700 ac., and the Pine Ridge District, 52,000 ac. Ph. 308/432-3367, 308/432-4475 (Pine Ridge), or 308/533-2257 (Bessey). Bird list available (about 250 spp., 36 nesters). Headquarters: 270 Pine St., Chadron, NE 69337. The Pine Ridge unit is largely native ponderosa pines with some grasslands, but the Bessey and McKelvie units consist mostly of Sandhills prairie with extensive conifer plantings. The Pine Ridge National Recreation area (6,600 ac.) has semiprimitive facilities (ph. 308/432-4475). Twenty Nebraska Sandhills prairie preserves or similar natural areas and lists of their associated floras and vertebrate faunas were summarized by Johnsgard (1995). Vertebrate lists for the Sandhills are also provided by Bleed and Flowerday (1989).

Significant mixed pine–grasslands areas in the Pine Ridge include Gilbert-Baker W.M.A., 2,337 ac., near Harrison, Sioux Co. (ph. 308/762-5605); Fort Robinson S.P., 22,000 ac. (ph. 308/665-2900), Peterson W.M.A., 2,640 ac. (ph. 308/762-5605), and Soldier Creek Wilderness, 7,800 ac. (ph. 308/432-4475), all near Crawford, Sioux Co. Sioux County Ranch (Guadalcanal Memorial Prairie), near Harrison, has 5,000 ac. of mostly short-grass prairie, managed by the Plains Prairie Resource Institute, Aurora, NE (ph. 402/694-5535). It also manages Bader Memorial Park (200 ac.) near Chapman, with tallgrass prairie, riparian woods, and wetlands. Lists of Bader Park birds (124 spp.), mammals (29 spp.), and herps (13 spp.) are available. An area of special paleontological importance is Ashfall Fossil Beds State Historical Park (360 ac.) in Antelope Co. (contact Park Manager, P.O. Box 66, Royal, NE 68773). For information on state parks, state recreation areas, and other state preserves, contact Nebraska Game and Parks Commission, Box 30370, Lincoln, NE 68503 (ph. 800/826-PARK). Over 400 Nebraska sites with public access were described by Johnsgard (2001b).

11. Niobrara S.P. 1,200 ac. Ph. 402/857-2273 or 402/857-3374. Located 25 miles above Gavin's Point Dam at the confluence of the Niobrara and Missouri Rivers, at the upper end of Lewis and Clark Lake. Bazille Creek W.M.A. (4,500 ac.) is nearby. Wetlands and riverine forests.

12. Niobrara Valley Preserve (T.N.C.). 60,000 ac. Ph. 402/722-4440. Near Springview, Keya Paha Co. Riparian transition zone between western ponderosa pine and eastern hardwood floodplain forest, including 26 miles of river frontage, with nearby sandhills prairie uplands. Lists of Niobrara Valley birds (Ducey 1989) and a biogeographic survey of plants and animals (Kaul, Kan-

tak, and Churchill 1988) are available. Other public-access T.N.C. preserves include the mixed-grass Willa Cather Prairie (609 ac.) near Red Cloud and a saline wetland site at Little Salt Fork Marsh (95 ac.) near Lincoln. For information on these and other T.N.C. holdings in Nebraska, contact their main field office, 1019 Leavenworth, Suite 100, Omaha, NE 68102 (ph. 402/342-0282).

13. Ogalala National Grassland. 94,344 ac. Ph. 308/432-3367 or 308/432-4475. Bird list of 302 spp. (covering all of Pine Ridge region) available from U.S. Forest Service, 270 Pine St., Chadron, NE 69337. Short-grass prairie and eroded badlands, including Toadstool Geological Park, an eroded badland area. Typical short-grass fauna, including two longspurs, burrowing owl, ferruginous hawk, and prairie dog.

14. Platte River crane-viewing sites. The Crane Meadows Nature Center, 240 ac. (ph. 308/382-1820), is located on the south side of I-80 exit 305, on Alda Road (admission fee). It offers exhibits, 7 miles of hiking trails, over 250 acres of prairie, programs, and daily (sunrise and sunset) guided trips to crane-viewing blinds between early March and early April (reservations needed; $15 fee; email: *info@cranemeadows.org*). See also the Lillian Annette Rowe sanctuary above. A nearby public-access crane-viewing platform overlooking the north shore of the main Platte channel is two miles south of I-80 exit 305. The Platte River Whooping Crane Maintenance Trust, 2,500 ac. (ph. 308/384-4633) is also nearby, located at 6611 Whooping Crane Drive, Wood River, NE 68883 (website: *www.whoopingcrane.org;* email: *trust@whoopingcrane.org*). This is a research facility, about 1 3/4 miles south and 1 mile east of I-80 exit 305. The ecology of the central Platte Valley was described by Johnsgard (1984) and by Currier, Lingle, and VanDerwalker (1985). At least 212 bird spp. have been recorded on trust lands (Faanes and Lingle 1995). Other vertebrates of the central Platte Valley include 53 mammals, 11 reptiles, and 7 amphibians (Johnsgard 1984). The Johnson No. 2 hydroplant (near Lexington) of the Central Nebraska Public Power and Irrigation District has an eagle-viewing site, open to individuals on weekends from 8 A.M. to 2 P.M., during the winter period while eagles are present (ph. 308/995-8601 or 308/324-2811).

15. Ponca S.P. 830 ac. Ph. 402/755-2284. Located along the Missouri River in northern Nebraska, beside one of the few unchanneled sections of the river. Mature hardwood forest, with 17 miles of hiking trails.

16. Rainwater Basin W.M.D. Two units, totaling 26,600 ac. Ph. 308/236-5015, 308/865-5310, or 308/385-6465. Bird list available (256 spp., 102 nesters), along with a suggested route map of 112 miles. Box 1786, Kearney, NE 68847. Shallow playa wetlands (84 publicly owned), surrounded by agricultural land. About 40 of these are designated as waterfowl production areas (W.P.A.), others are wildlife management areas (W.M.A.). The largest number of W.P.A.'s are in Clay Co. (11), followed by Kearney Co. (9), and Fillmore Co. (8). Among the best for aquatic birds are Funk W.P.A. (near Funk; observation sites),

Massie W.P.A. (near Clay Center; observation tower), and Harvard W.P.A. (near Harvard). Some shallow or temporary lagoons such as Meadowlark (previously Sandpiper) W.P.A. (near Sutton) are notable for shorebirds. Roadside viewing is most feasible at Mallard Haven and Eckhardt W.P.A.'s (both near Shickley) and at Pintail W.P.A. (near Aurora). The wetlands may support 7 to 9 million ducks and 3 to 5 million geese in spring, especially white-fronted geese, snow geese, mallards, and pintails. Further information is available at *www.ngpc.state.ne.us/wildlife/rwbjv/rwbjv1/html*.

17. Scotts Bluff National Monument. 2,988 ac. Ph. 308/436-4340. P.O. Box 427, Gering, NE 69341. Conifer-covered bluffs surrounded by short-grass uplands. The monument's terrestrial vertebrates (98 birds, 28 mammals, and 12 herptiles) were documented by Cox and Franklin (1989).

18. Schramm S.P. 330 ac. Ph. 402/332-3901. Adjacent to the lower Platte River, with excellent hardwood forests. Includes an aquarium and outdoor education center (admission fee). One of the few breeding sites known in Nebraska for the summer tanager and Kentucky warbler.

19. Valentine N.W.R. 71,516 ac. Ph. 402/376-3789. Bird and mammal lists available (270 bird spp., 95 nesters). Contact Ft. Niobrara N.W.R., Hidden Timber Rt., Valentine, NE 69201. Marshes, shallow lakes, wet meadows, and sandhills prairie. The breeding birds are similar to those of Crescent Lake, but the marshes and lakes are generally larger, deeper, and less alkaline.

20. Wildcat Hills S.R.A. 935 ac. Ph. 308/436-3777 or 308/436-3394. An area of mixed ponderosa pine forest, prairies, and brushlands in ridge and canyon topography. The Wildcat Hills Nature Center (Nebraska Game and Parks facility) offers close observations of coniferous forest birds such as red crossbills and pygmy nuthatches at bird feeders. Poor-wills are common in summer. Nearby Buffalo Creek W.M.A. (2,880 ac.) has similar habitats (ph. 308/436-6888).

Kansas (see also Gress and Potts 1993)

1. Cimarron National Grassland (N.G.) 108,175 ac. Ph. 620/697-4621. 737 Villymaca, Elkhart, KS 67950. Bird list (includes adjacent Comanche N.G.) available (345 spp., 72 nesters); Cable, Seltman, and Cook (1997) published an annotated list of 342 spp. for Cimarron N.G. and adjacent Morton Co., Kansas. Short-grass plains and sandsage steppe.

2. Cheyenne Bottoms Waterfowl Management Area. 18,000 ac. Ph. 620/665-0231. Bird list available (319 spp., 104 nesters). Rt. 1, Great Bend, KS 67530. Marshes and moist mixed-grass prairie, with the water levels controlled by dikes. Zimmerman (1993) has described this internationally important wetland. The bird list of more than 300 spp. is one of the largest in the entire Great Plains.

3. El Dorado S.P. 8,000 ac. Ph. 620/321-7180. Near El Dorado. Includes prairie remnants. Other prairies on private land are located between El Dorado, Elmdale, and Cottonwood Falls.

4. Flint Hills N.W.R. 18,463 ac. Ph. 620/392-5553. Bird list available (294 spp., 90 nesters). Box 128, Hartford, KS 66854. Tallgrass and mixed-grass prairie, hardwood forests, and shallow wetlands as well as part of John Redmond Reservoir. Waterfowl, gulls, shorebirds, and herons are numerous during migration.

5. Kirwin N.W.R. 10,778 ac. Ph. 785/543-6673. Bird list available (205 spp., 46 nesters). Kirwin, KS 67644. Mixed-grass prairie and hardwoods surrounding the 5,000 ac. Kirwin Reservoir. Large numbers of migratory waterfowl stop here, and bald eagles overwinter regularly.

6. Konza Prairie. 8,616 ac. Ph. 785/272-5115. Near Manhattan. Major research grassland (tallgrass and mixed-grass); operated by Kansas State University Division of Biology. See Zimmerman (1993) and Reichman (1987). Zimmerman (1993) published an annotated bird list of 207 spp., and there is a mammal list of 40 spp. (McMillan et al. 1997). Over 500 plant spp. have been tallied, plus 9 reptiles and 25 amphibians. The Nature Conservancy also owns 2,818 ac. of tallgrass prairie near Cassoday. Tours can be arranged by contacting the T.N.C. office, Box HP, S.E. Quincy, #301, Topeka, KS 66612 (ph. 785/233-4400). The Conservancy also owns the Smoky Valley Ranch, 16,320 ac. of shortgrass prairie, chalk bluffs, and rocky ravines, in Logan Co. Access is 25 miles south of Oakley and I-70. For information, contact the T.N.C. office in Topeka.

Zimmerman and Patti (1988) have described 24 prairie sites desirable for birding in Kansas, as well as 18 forest or forest-prairie mosaic sites, plus additional sites in western Missouri. Major state-owned wildlife areas (W.A.) include Berentz/Dick W.A., 1,360 ac., and Elk City W.A., 12,240 ac., both near Independence (Montgomery Co.); Fall River W.A., 8,392 ac., near Eureka, (Greenwood Co.); John Redmond W.A., 1,472 ac., near Burlington (Coffey Co.); Melvern W.A., 9,407 ac., near Lebo (Osage Co.); Perry W.A., 10,984 ac., near Valley Falls (Jefferson Co.); Toronto W.A., 4,766 ac., and Woodson W.A., 3,065 ac., both near Toronto (Woodson Co.). Contact Kansas Department of Wildlife and Parks, 900 S.W. Jackson St., Topeka, KS 66612 (ph. 785/296-2281).

Many state parks also offer excellent nature viewing areas. In western Kansas, there is Cedar Bluff S.P., near Hays, with about 9,000 ac. of wildlife habitat (ph. 785/726-3212). Near Norton are Prairie Dog S.P. and Norton W.A. (6,500 ac.), supporting a good prairie dog population and associated raptors (ph. 785/877-2953). Scott S.P., near Scott City, has about 1,100 ac. of wildlife habitat, with a small spring-fed lake and canyon and bluff topography (ph. 620/872-2061). It also has small herds of elk and bison. In south-central Kansas, Cheney S.P. is close to Wichita, with 5,250 ac. of natural habitat (ph. 316/542-3664). The Great Plains Nature Center, 6232 E. 29th St. N., Wichita, KS 67220 (ph. 316/683-5499), is an excellent source of nature information and has both indoor and outdoor exhibits (240 ac.). North of Wichita, near McPherson, is Maxwell W.A., 2,500 ac., supporting managed herds of elk and bison (ph. 316/767-5900). East of Wichita is El Dorado S.P., with 4,600 ac. of wildlife habitat (ph. 316/321-7180). A preserved area of sandsage prairie, wetlands, and woods is in Sandhills S.P., 1,123 ac., near Hutchinson (ph. 316/542-3664). Nearby to the north are the McPherson Valley wetlands (1,310 ac.), reclaimed wetlands managed by the Kansas Department of Wildlife and Parks (ph. 316/766-5900). Near Salina is Kanopolis S.P., with 12,500 ac. of wildlife habitat, with many sandstone canyons and bluffs (ph. 785/546-2565). Near Russell is Wilson S.P., 927 ac., in the Smoky Hills region, with some native prairie (ph. 785/658-2465). In eastern Kansas near Lawrence is Clinton S.P., with 1,500 ac. of native prairie and woods (ph. 785/842-8562). In the same region are also Eisenhower S.P. (ph. 785/528-4102) and Pomona S.P., with 490 ac. of prairies and woods (ph. 785/828-4933). North of Lawrence is Perry S.P., including about 1,000 ac. of marshes (ph. 785/246-3449).

7. Marais des Cygnes N.W.R. 7,500 ac. Ph. 913/352-8956. Near Pleasanton; mostly bottomland hardwood forest. Bird list (321 spp., 111 nesters) available. 24141 Kansas Hwy 52, Pleasanton, KS 66075.

8. Quivira N.W.R. 21,820 ac. Ph. 316/486-2393. Bird list available (252 spp., 88 nesters). Box G, Stafford, KS 67578. Mature hardwoods, sandhill grasslands, rangeland, farmland, and alkaline marshes. An important breeding site for snowy plovers; also used by waterfowl and whooping and sandhill cranes during migration.

9. The Tallgrass Prairie National Preserve (about 10,000 ac.) is located north of Cottonwood Falls in the heart of the Flint Hills. Much of this preserve is used for cattle grazing, but a small area is open to the public. Contact the Tallgrass Prairie National Preserve, P.O. Box 585, Cottonwood Falls, KS 66845 (ph. 316/273-6034 or 273-8494).

Oklahoma (see also Oklahoma Department of Conservation 1994)

1. Black Kettle National Grassland. 32,000 ac. Ph. 580/497-2143. P.O. Box 266, Cheyenne, OK 73628. Mixed-grass and tallgrass prairie and upland woods. Extends into Texas panhandle.

2. Chickasaw National Recreation Area. 10,000 ac. Ph. 580/622-3161. Hardwood forests around the Lake of the Arbuckles, with prairies on slopes and hilltops. Bird list (160 spp.) available. P.O. Box 201, Sulfur, OK 73086.

3. Deep Fork N.W.R. Ph. 918/756-0815. A newly established refuge near Okmulgee of bottomland hardwoods and wetlands along Deep Fork River. Bird list (254 spp.) available. P.O. Box 816, Okmulgee, OK 74447.

4. Little River N.W.R. 15,000 ac. Ph. 405/584-6211. P.O. Box 340, Broken Bow, OK 74728. Consists of bottomland hardwoods along the Little River. A bird list of 254 spp. is available, including such unique southeastern species as Swainson's warbler and anhinga. Nearby are Ouachita National Forest, 1,585,000 ac., mostly in Arkansas, and McCurtain County Wilderness Area, 15,220 ac. of virgin oak–pine forest in north-central McCurtain Co. (ph. 580/241-7875).

5. Optima N.W.R. 4,333 ac. Ph. 580/664-2205. Bird list available (246 spp., 106 nesters). Near Guymon. R.R. 1, Box 68, Butler, OK 73625. Short-grass and mixed-grass prairie, sage-dominated shrubsteppe, and hardwoods adjoining Optima Reservoir on the Beaver River.

6. Ozark Plateau N.W.R. 2,215 ac. Ph. 918/773-5251. Includes several bat caves and surrounding forest habitat. Rt. 1, Box 18A, Vian, OK 74962.

7. Salt Plains N.W.R. 32,000 ac. Ph. 580/626-4794. Mostly consists of Salt Plains Reservoir, but with salt flats that provide nesting habitat for snowy plover and other shorebirds. A bird checklist of 243 spp. (98 nesters) is available. Rt. 1, Box 76, Jet, OK 73749.

8. Sequoyah N.W.R. 20,800 ac. Ph. 918/773-5251. Bird list available (256 spp., 96 nesters). Rt. 1, Box 18A, Vian, OK 74962. Adjoins Kerr Reservoir, at the confluence of the Canadian and Arkansas Rivers. Grasslands, hardwoods, and wetlands. To the south, within the Ouachita National Forest (200,000 ac., ph. 918/653-2991), is the Winding Stair Mountain National Recreational Area of more than 70,000 ac., including Beech Creek Scenic and Botanical Area, Black Fork Mountain Wilderness, Indian Nations Scenic and Wildlife Area, and Upper Kiamichi River Wilderness. Along the northern border of Ouachita National Forest is Little River N.W.R. (see above).

9. Tallgrass Prairie Preserve. 44,503 ac. Ph. 918/287-4803 (preserve headquarters) or 918/585-1117 (T.N.C. field office). Near Pawhuska, Osage Co. Tallgrass prairie and about a thousand reintroduced bison; the largest area of preserved tallgrass prairie in the United States. A bird list for the tallgrass prairie area and wooded Osage Hills has 266 spp., with 97 known nesters (contact Oklahoma Department of Conservation, 1801 N. Lincoln Blvd., Oklahoma City, OK 73105; ph. 405/521-3855). Nearby is the Prairie National Wild Horse Preserve (18,000 ac., ph. 918/333-5575).

10. The Nature Conservancy (T.N.C.) sites. A short-grass preserve (Black Mesa Nature Preserve) of 1,600 ac. is located at Black Mesa, Cimarron Co. Ph. 580/426-2222. Black Mesa State Park (ph. 580/521-2409) is contiguous and also includes short-grass prairie. The Black Mesa region supports several unique species, including the juniper titmouse and common raven. The Pontotoc Ridge Nature Preserve (2,900 ac.), Pontotoc Co., near Ada (ph. 918/ 580-2224), includes tallgrass and short-grass prairies as well as bottomland forest located at the eastern edge of the Arbuckle Uplift. The Arkansas River Least Tern Preserve (1,400 ac.), located in North Tulsa, not only protects nesting least terns but also migrating shorebirds along 5 miles of river frontage. The Oxley Nature Center is nearby. Other smaller T.N.C. preserves in Oklahoma also include remnant prairies; for information, contact their field office, 23 West Fourth, #200, Tulsa, OK 74103 (ph. 918/585-1117).

11. Tishomingo N.W.R. 16,464 ac. Ph. 580/371-2402. Bird list available (252 spp., 81 nesters). Rt. 1, Box 152, Tishomingo, OK 73460. Grasslands, wetlands,

oak-hickory woods, and cultivated lands adjoining Lake Texoma reservoir of the Red River.

12. Washita N.W.R. 8,200 ac. Ph. 580/664-2205. Bird and mammal lists available (229 bird spp., 67 nesters). Rt. 1, Box 68, Butler, OK 73625. Grasslands (mostly short-grass), wetlands, woods, and cultivated lands adjoining Foss Reservoir of the Washita River.

13. Wichita Mountains N.W.R. 59,020 ac. Ph. 580/429-3221. Bird list available (278 spp., 61 nesters). Rt. 1, Box 448, Indiahoma, OK 73552. Grassy and oak-covered uplands that are the eroded remnants of ancient mountains, and more than a hundred small wetlands, mostly impoundments. Managed partly for bison, elk, and longhorn cattle.

14. Wildlife management areas and state parks. Oklahoma has over 60 wildlife management areas and nearly 50 state parks. Information on the wildlife management areas can be obtained from the Oklahoma Department of Wildlife Conservation's website: *www.wildlifedepartment.com*. State park information is available from the Oklahoma Tourism and Recreation Department, 15 N. Robinson, P.O. Box 52002, Oklahoma City, OK 73152-2002 (ph. 405/521-2406).

Texas Panhandle

1. Buffalo Lake N.W.R. 7,700 ac. Ph. 806/499-3382. A bird checklist of 344 spp. is available. Entry fee. P.O. Box 228, Umbarger, TX 79091. An impounded creek, with about 1,000 ac. of surface water and adjoining grasslands. A major wintering area for sandhill cranes and waterfowl.

2. Rita Blanca National Grassland. 77,414 ac. partly in Oklahoma and contiguous with Kiowa National Grassland in New Mexico. Ph. 806/562-4254. Short-grass prairie, with pronghorns, prairie dogs, and other plains species. Box 38, Texline, TX 79087. Nearby is Lake Rita Blanca S.P. (1,668 ac.), which is administered by Palo Duro Canyon S.P. (see below). Lake Rita Blanca is a major waterfowl wintering area. Not far to the south is Lake Meredith National Recreation Area (44,951 ac.), a reservoir on the Canadian River (ph. 806/857-3151).

3. State parks. Major state parks in the Texas panhandle include the spectacularly deep Palo Duro Canyon S.P., Rt. 2, P.O. Box 285, Canon, TX 79015 (ph. 806/488-2227). There is also Caprock Canyon S.P., P.O. Box 204, Quitaque, TX 79255 (ph. 806/455-1492). Seyffert (2001) has documented the birds of the panhandle region. For information, contact the Texas Parks and Wildlife Department, 4200 Smith School Rd., Austin, TX 78744 (ph. 512/389-4800, or 800/792-1111 if out of state).

APPENDIX 3

Birds of the Great Plains States

These lists include approximately 300 breeding and more than 100 transient species that rather regularly occur from North Dakota through Oklahoma plus the Texas panhandle. They exclude nearly all extinct, regionally extirpated, and very rare species. Ranges and habitats indicated for the breeding species refer to the nesting season only; nonbreeders often occur over much wider regions and in different habitats. The indicated regional status for breeding species is largely based on recent state atlases, including South Dakota's (Peterson 1995), Nebraska's (Mollhoff 2001), and Kansas's (Busby and Zimmerman 2001). References to Texas status pertain only to the panhandle and are based on the still unpublished but online *Texas Breeding Bird Atlas (http://tbba.cbi.tamucc.edu/)*. Oklahoma's breeding bird atlas is still in preparation, and no North Dakota breeding summary has appeared since Stewart's (1975).

Permanent residents and those species that regularly to uncommonly overwinter within this defined region are so identified; the others usually migrate and winter beyond the region's limits. "Declining nationally" and "Increasing nationally" refer to those species that declined or increased surveywide at a significant rate (P < 0.1%) during North American breeding bird surveys, 1966–2000 *(http://www.mbr-pwrc.usgs.gov/bbs/)*. "Probably declining" and "Probably increasing" refer to species that have nationally declined or increased at least 1 percent annually during that period, but not at statistically significant levels. "Regionally increasing" and "Regionally declining" refer to probable but not always statistically significant trends in the Great Plains states or the Central States Region of the breeding bird survey database. Population trends are not indicated for transient species. Family sequence as well as vernacular and Latin names follow the seventh edition of the American Ornithologists' Union checklist (1998). Species considered by the author (Johnsgard 2001a) to be especially characteristic of the Great Plains grasslands are underlined.

Family Gaviidae—Loons

Common Loon. *Gavia immer.* Northeastern North Dakota. Larger lakes; wintering irregularly southward. Increasing nationally.

Family Podicipedidae—Grebes

Pied-billed Grebe. *Podilymbus podiceps.* Entire region. Marshes; winters commonly southward. Increasing nationally.

Horned Grebe. *Podiceps auritus.* North Dakota and northern South Dakota. Ponds, shallow lakes, and marshes; winters uncommonly southward. Declining nationally.

Red-necked Grebe. *Podiceps grisegena.* North Dakota and (rarely) eastern South Dakota. Medium to larger shallow lakes.

Eared Grebe. *Podiceps nigricollis.* North Dakota to central Nebraska; rare in Kansas. Ponds, shallow lakes, marshes. Increasing nationally.

Western Grebe. *Aechmophorus occidentalis.* North Dakota to central Nebraska; rare in Kansas. Shallow lakes, large marshes.

Clark's Grebe. *Aechmophorus clarkii.* North Dakota to western Nebraska. Shallow lakes, large marshes.

Family Pelecanidae—Pelicans

American White Pelican. *Pelecanus erythrorhynchos.* North Dakota and South Dakota. Lakes with low islands. Probably increasing nationally.

Family Phalacrocoracidae—Cormorants

Double-crested Cormorant. *Phalacrocorax auritus.* North Dakota to Nebraska, locally to Oklahoma. Lakes and marshes. Increasing nationally and regionally.

Family Anhingidae—Anhingas

Anhinga. *Anhinga anhinga.* Southeastern Oklahoma, where rare. Swamps.

Family Ardeidae—Bitterns and Herons

American Bittern. *Botaurus lentiginosus.* Entire region, but very local. Marshes.

Least Bittern. *Ixobrychus exilis.* Eastern North Dakota to central Oklahoma. Marshes. Probably declining nationally.

Great Blue Heron. *Ardea herodias.* Entire region. Marshes, riparian woods; winters commonly southward. Increasing nationally.

Great Egret. *Ardea alba.* Eastern and central Kansas to Oklahoma. Marshes, wet forests. Increasing nationally and regionally.

Snowy Egret. *Egretta thula.* Oklahoma, sporadically north to North Dakota. Marshes.

Little Blue Heron. *Egretta caerulea.* Oklahoma, north rarely to North Dakota. Marshes. Declining nationally.

Cattle Egret. *Bubulcus ibis.* Kansas to Oklahoma and Texas, sporadically north to North Dakota. Pastures, marshes. Increasing regionally.

Green Heron. *Butorides virescens.* Eastern North Dakota to central Oklahoma. Wooded ponds, marshes. Declining nationally.

Black-crowned Night-heron. *Nycticorax nycticorax.* North Dakota to Oklahoma and Texas. Wooded ponds, marshes. Increasing nationally.

Yellow-crowned Night-heron. *Nyctanassa violacea.* East-central Kansas and Oklahoma, sporadically northward. Wooded marshes, lakes. Probably declining nationally.

Family Threskiornithidae—Ibises and Spoonbills

White Ibis. *Eudocimus albus.* Southeastern Oklahoma. Swamps.

White-faced Ibis. *Plegadis chihi.* Irregularly breeds throughout region, mainly to south. Marshes. Candidate for federal listing as threatened or endangered. Increasing nationally, probably also regionally.

Family Cathartidae—American Vultures

Black Vulture. *Coragyps atratus.* Southeastern Oklahoma. Diverse open habitats.

Turkey Vulture. *Cathartes aura.* North Dakota to Oklahoma and Texas. Diverse open country habitats. Increasing nationally and regionally.

Family Anatidae—Swans, Geese, and Ducks

Canada Goose. *Branta canadensis.* North Dakota to Kansas, locally farther south. Lakes, marshes, parks; winters widely throughout region. Increasing nationally and regionally.

Trumpeter Swan. *Cygnus buccinator.* Reintroduced in South Dakota; now also local in Nebraska Sandhills. Large marshes; winters near breeding grounds. Slowly increasing in northern Great Plains.

Wood Duck. *Aix sponsa.* Central North Dakota south to central Oklahoma, locally to Texas. Riparian forests, swamps; winters uncommonly southward. Increasing nationally and regionally.

Gadwall. *Anas strepera.* North Dakota to Kansas, locally farther south. Marshes; winters commonly southward. Increasing nationally and regionally.

American Wigeon. *Anas americana.* North Dakota to western Nebraska. Marshes; winters commonly southward. Probably increasing nationally.

American Black Duck. *Anas rubripes.* North Dakota, where rather rare; accidental in South Dakota. Marshes. Probably declining nationally.

Mallard. *Anas platyrhynchos.* North Dakota to western Oklahoma. Marshes; winters widely and commonly in region. Increasing nationally.

Mottled Duck. *Anas fulvigula.* Rare breeder north to Kansas. Marshes. Declining nationally.

Blue-winged Teal. *Anas discors.* North Dakota to western Oklahoma and Texas. Marshes.

Cinnamon Teal. *Anas cyanoptera.* Local from western North Dakota to Texas. Marshes.

Northern Shoveler. *Anas clypeata.* North Dakota to western Oklahoma and Texas. Marshes; winters uncommonly in region. Increasing nationally and regionally.

Northern Pintail. *Anas acuta.* North Dakota to western Oklahoma and Texas. Marshes; winters commonly in region. Declining nationally.

Green-winged Teal. *Anas crecca.* North Dakota to Kansas, locally farther south. Marshes; winters commonly in region. Increasing nationally.

Canvasback. *Aythya valisineria.* North Dakota to central Nebraska. Larger marshes; winters commonly in region.

Redhead. *Aythya americana.* North Dakota to central Nebraska, locally farther south. Larger marshes; winters uncommonly in region. Increasing nationally.

Ring-necked Duck. *Aythya collaris.* North Dakota to eastern South Dakota. Boggy marshes; winters commonly in region. Increasing nationally.

Lesser Scaup. *Aythya affinis.* North Dakota to northern South Dakota, locally farther south. Small lakes; winters uncommonly in region.

Bufflehead. *Bucephala albeola.* Northern North Dakota, in Turtle Mountains; one South Dakota breeding. Wooded lakes; winters commonly in region.

Common Goldeneye. *Bucephala clangula.* Northern North Dakota, in Turtle Mountains. Wooded lakes; winters commonly in region.

Hooded Merganser. *Lophodytes cucullatus.* Northern North Dakota, locally south to at least Kansas. Wooded lakes and streams; winters commonly in region.

Common Merganser. *Mergus merganser.* Rare from North Dakota to Nebraska. Deep, clear lakes; winters uncommonly in region. Increasing nationally.

Ruddy Duck. *Oxyura jamaicensis.* North Dakota to Nebraska, locally farther south. Marshes; winters uncommonly in region. Probably increasing nationally and regionally.

Family Accipitridae—Kites, Hawks, Eagles, and Allies
Osprey. *Pandion haliaetus.* Rare in South Dakota (state-threatened) and North Dakota (attempted nestings). Rivers and lakes. Increasing nationally, but still rare regionally.

Mississippi Kite. *Ictinia mississippiensis.* Texas and Oklahoma north to Kansas and (very locally) western Nebraska. Open woods, suburbs.

Bald Eagle. *Haliaeetus leucocephalus.* Local from North Dakota south to Kansas; nationally threatened. Rivers and lakes with large trees; winters commonly and widely in region. Increasing nationally and regionally.

Northern Harrier. *Circus cyaneus.* North Dakota to Oklahoma and Texas. Marshes; winters commonly in region. Declining regionally.

Sharp-shinned Hawk. *Accipiter striatus.* North Dakota to eastern Nebraska, locally farther south. Mature forests; winters commonly in region. Increasing nationally.

Cooper's Hawk. *Accipiter cooperii.* North Dakota to Oklahoma and Texas, mainly eastwardly. Mature forests, locally in urban areas; winters uncommonly in region. Increasing nationally.

Northern Goshawk. *Accipiter gentilis.* Black Hills. Coniferous forest. Candidate for federal listing as threatened or endangered. Probably declining nationally.

Red-shouldered Hawk. *Buteo lineatus.* Eastern Kansas to eastern Oklahoma. Moist forests. Increasing nationally.

Broad-winged Hawk. *Buteo platypterus.* North Dakota, local in South Dakota, also eastern Kansas (rarely) and eastern Oklahoma. Deciduous or mixed forests. Declining regionally.

Swainson's Hawk. *Buteo swainsoni.* North Dakota to Oklahoma and Texas, mainly westward. Grasslands with scattered trees.

Red-tailed Hawk. *Buteo jamaicensis.* Entire region. Open areas with tall trees; winters throughout region. Increasing nationally and regionally.

Ferruginous Hawk. *Buteo regalis.* North Dakota to western Oklahoma and Texas, mainly westward. Arid grasslands; winters uncommonly in region. Candidate for federal listing as threatened or endangered. Increasing nationally.

Golden Eagle. *Aquila chrysaetos.* Western North Dakota to western Oklahoma and Texas. Widespread but rare in Great Plains, winters throughout region. Probably increasing nationally.

Family Falconidae—Falcons

American Kestrel. *Falco sparverius.* Entire region; largely resident. Diverse habitats.

Merlin. *Falco columbarius.* Northern North Dakota. Rare in cities, largely resident but also winters southward. Increasing nationally.

Peregrine Falcon. *Falco peregrinus.* Rare and local; usually from released birds. Historically bred in the Black Hills and Nebraska's Pine Ridge. Mostly migratory throughout region. Increasing nationally; recent urban breedings in Nebraska and North Dakota.

Prairie Falcon. *Falco mexicanus.* Western North Dakota to extreme western Oklahoma and Texas. Open country, near cliffs; winters uncommonly in region. Increasing nationally.

Family Phasianidae—Partridges, Grouse, and Turkeys

Gray Partridge. *Perdix perdix.* North Dakota and northern South Dakota, rarely to Nebraska. Farmlands and weedy fields (introduced species), resident.

235

Ring-necked Pheasant. *Phasianus colchicus*. North Dakota to northern Oklahoma and Texas. Farmlands and weedy fields (introduced species), resident. Declining nationally.

Ruffed Grouse. *Bonasa umbellus*. Black Hills; reintroduced into eastern Kansas. Aspen-conifer woods, resident. Probably declining nationally.

Greater Sage-grouse. *Centrocercus urophasianus*. Southwestern North Dakota and western South Dakota; extirpated from Nebraska, resident. Sage scrub. Regionally declining.

Sharp-tailed Grouse. *Tympanuchus phasianellus*. North Dakota south to central Nebraska, previously to northern Kansas. Mixed or short grasslands, resident. The western race *columbianus* is a candidate for federal listing; the Great Plains population *jamesi* is stable.

Greater Prairie-chicken. *Tympanuchus cupido*. Eastern North Dakota, central South Dakota, central Nebraska, eastern Kansas, and northeastern Oklahoma. Tallgrass prairies with nearby croplands, resident. Regionally declining.

Lesser Prairie-chicken. *Tympanuchus pallidicinctus*. Southwestern Kansas, western Oklahoma, and Texas. Arid grassland with shrubs, resident. Regionally declining; candidate for federal listing as threatened or endangered.

Wild Turkey. *Meleagris gallopavo*. Entire region. Wooded habitats, resident. Increasing nationally.

Family Odontophoridae—New World Quail

Scaled Quail. *Callipepla squamata*. Southwestern Kansas, western Oklahoma, and Texas. Arid grasslands, resident. Declining nationally.

Northern Bobwhite. *Colinus virginianus*. Central South Dakota south to Oklahoma and Texas. Woodland edges, resident. Declining nationally.

Family Rallidae—Rails, Gallinules, and Coots

Yellow Rail. *Coturnicops noveboracensis*. Eastern North Dakota. Wet meadows.

Black Rail. *Laterallus jamaicensis*. Central Kansas, very rare in Nebraska. Wet meadows.

King Rail. *Rallus elegans*. Southeastern North Dakota to central Oklahoma. Marshes.

Virginia Rail. *Rallus limicola*. North Dakota to southern Oklahoma. Marshes. Increasing nationally.

Sora. *Porzana carolina*. North Dakota to central Kansas. Marshes. Increasing nationally.

Purple Gallinule. *Porphyrula martinica*. Eastern and southern Oklahoma. Marshes.

Common Moorhen. *Gallinula chloropus*. Eastern Kansas south to central Oklahoma. Rare in Nebraska, one South Dakota record. Marshes. Probably increasing nationally.

American Coot. *Fulica americana.* Entire region; winters commonly toward south. Marshes.

Family Gruidae—Cranes
Sandhill Crane. *Grus canadensis.* Once extirpated; now very rare in Nebraska and North Dakota. Meadows, marshes, shallow lakes; winters commonly in south. Increasing nationally; Platte Valley is critical spring staging area.
Whooping Crane. *Grus americana.* Extirpated as a regional breeder (previously bred south to Nebraska). Nationally endangered; central Platte Valley is considered as critical migration habitat.

Family Charadriidae—Plovers
Snowy Plover. *Charadrius alexandrinus.* Central Kansas south to Oklahoma and Texas. Sandy beaches. Candidate for federal listing as threatened or endangered.
Piping Plover. *Charadrius melodus.* North Dakota south to central Nebraska and eastern Kansas. Sandy beaches of rivers and lakes. Interior population federally threatened in Great Plains.
Killdeer. *Charadrius vociferus.* Entire region. Gravelly fields, pastures; winters uncommonly in southern regions. Declining nationally.
Mountain Plover. *Charadrius montanus.* Western edges of Nebraska (where threatened), Kansas, Oklahoma, and Texas. Arid plains. Candidate for federal listing as threatened or endangered.

Family Recurvirostridae—Stilts and Avocets
Black-necked Stilt. *Himantopus mexicanus.* Sporadic or local in central and southern parts of region, north locally to Nebraska and the Dakotas. Marshes, especially alkaline wetlands. Probably declining regionally.
American Avocet. *Recurvirostra americana.* North Dakota to central Kansas, western Oklahoma, and Texas. Marshes, especially alkaline wetlands.

Family Scolopacidae—Sandpipers and Phalaropes
Willet. *Catoptrophorus semipalmatus.* North Dakota south to central Nebraska. Marshes.
Spotted Sandpiper. *Actitis macularia.* North Dakota south to western Oklahoma and Texas. Widespread near water.
Upland Sandpiper. *Bartramia longicauda.* North Dakota south to central Oklahoma and Texas. Native grasslands and meadows. Increasing nationally and regionally.
Long-billed Curlew. *Numenius americanus.* Western North Dakota south through central Nebraska, southwestern Kansas, western Oklahoma, and Texas. Mixed-grass prairies and meadows. Declining nationally and regionally; candidate for federal listing as threatened or endangered.

Marbled Godwit. *Limosa fedoa.* North Dakota south to eastern South Dakota. Mixed-grass prairies.

Common Snipe. *Gallinago gallinago.* North Dakota south locally to central Nebraska. Marshes, fens, wet meadows; winters uncommonly in southern regions.

American Woodcock. *Scolopax minor.* Eastern North Dakota south to eastern Oklahoma. Moist woodlands.

Wilson's Phalarope. *Phalaropus tricolor.* North Dakota to western Kansas. Marshes, especially alkaline wetlands. Probably declining nationally.

Family Laridae—Gulls and Terns

Franklin's Gull. *Larus pipixcan.* North Dakota south to eastern South Dakota; rarely to Kansas. Prairie marshes. Probably increasing nationally.

Ring-billed Gull. *Larus delawarensis.* North Dakota; local in eastern South Dakota. Larger lakes with islands; winters commonly in region. Increasing nationally.

California Gull. *Larus californicus.* Western North Dakota and northeastern South Dakota; expanding range. Larger marshes and lakes with bare islands.

Common Tern. *Sterna hirundo.* North Dakota and eastern South Dakota. Lakes. Declining nationally.

Forster's Tern. *Sterna forsteri.* North Dakota south to eastern South Dakota, locally south to Kansas. Marshes.

Least Tern. *Sterna antillarum.* River drainages from North Dakota to southern Oklahoma. Interior race *athalassos* is nationally endangered. Sandy or gravelly beaches.

Black Tern. *Chlidonias niger.* North Dakota south to west-central Kansas. Marshes. Candidate for federal listing as threatened or endangered. Probably declining nationally.

Family Columbidae—Pigeons and Doves

Rock Dove. *Columba livia.* Entire region; introduced. Cities and farms, resident.

Eurasian Collared-dove. *Streptopelia decaocto.* Introduced; local but expanding northward, rarely to North Dakota. Cities and towns, resident. Increasing nationally.

Mourning Dove. *Zenaida macroura.* Entire region; winters commonly in region. Diverse habitats. Declining nationally.

Family Cuculidae—Cuckoos and Anis

Black-billed Cuckoo. *Coccyzus erythropthalmus.* North Dakota south to Oklahoma and Texas. Deciduous or mixed woods. Declining nationally.

Yellow-billed Cuckoo. *Coccyzus americanus.* South Dakota south to central Oklahoma. Riparian woods or deciduous forest. Declining nationally and regionally; the western race *occidentalis* is candidate for federal listing as threatened or endangered.

Greater Roadrunner. *Geococcyx californianus.* Southern Kansas to Oklahoma and Texas. Arid scrub, resident.

Family Tytonidae—Barn Owls

Barn Owl. *Tyto alba.* South Dakota south to Oklahoma and Texas. Diverse habitats, resident. Probably declining nationally.

Family Strigidae—Typical Owls

Eastern Screech-owl. *Otus asio.* Entire region, mainly eastward. Deciduous woods, resident. Probably increasing nationally.

Great Horned Owl. *Bubo virginianus.* Entire region. Diverse habitats, resident.

Burrowing Owl. *Athene cunicularia.* North Dakota south to Oklahoma and Texas, mainly westward. Short-grass areas, especially near prairie dog towns; winters uncommonly in southern regions. Probably declining regionally; candidate for federal listing as threatened or endangered.

Barred Owl. *Strix varia.* Eastern edges of states from North Dakota south, becoming more widespread in Kansas and Oklahoma. Deciduous forests, resident. Increasing nationally.

Long-eared Owl. *Asio otus.* Entire region, but local. Deciduous forests, resident.

Short-eared Owl. *Asio flammeus.* North Dakota south to Kansas, locally to Oklahoma. Grasslands, meadows, and marshes, resident. Declining nationally.

Northern Saw-whet Owl. *Aegolius acadicus.* Black Hills, scattered nestings elsewhere; winters uncommonly in region. Coniferous or mixed woods.

Family Caprimulgidae—Goatsuckers

Common Nighthawk. *Chordeiles minor.* Entire region. Cities, open grasslands. Declining nationally.

Common Poorwill. *Phalaenoptilus nuttallii.* Western North Dakota to western Nebraska, the Flint Hills of Kansas, and western Oklahoma and Texas. Arid, rocky scrub. Probably increasing nationally.

Chuck-will's-widow. *Caprimulgus carolinensis.* Eastern Nebraska south through eastern and central Kansas to central Oklahoma. Deciduous woods. Declining nationally.

Whip-poor-will. *Caprimulgus vociferus.* Southeastern South Dakota through eastern Nebraska and Kansas to east-central Oklahoma. Open woods. Declining nationally.

Family Apodidae—Swifts

Chimney Swift. *Chaetura pelagica.* Entire region, but rarer westward. Cities. Declining nationally.

White-throated Swift. *Aeronautes saxatalis.* Western South Dakota and western Nebraska. Cliffs and canyons. Declining nationally.

Family Trochilidae—Hummingbirds

Ruby-throated Hummingbird. *Archilochus colubris.* Eastern parts of entire region, extending west to central Oklahoma. Deciduous woods. Increasing nationally.

Broad-tailed Hummingbird. *Selasphorus platycercus.* Extreme western Oklahoma, previously also Black Hills. Open woods.

Family Alcedinidae—Kingfishers

Belted Kingfisher. *Ceryle alcyon.* Entire region. Streams and lakes; winters uncommonly in region. Declining nationally.

Family Picidae—Woodpeckers

Lewis's Woodpecker. *Melanerpes lewis.* Western South Dakota and northwestern Nebraska. Logged or burned coniferous forest.

Red-headed Woodpecker. *Melanerpes erythrocephalus.* Entire region. Open deciduous woods; winters uncommonly in region. Declining nationally.

Golden-fronted Woodpecker. *Melanerpes aurifrons.* Texas and southwestern Oklahoma. Open woods, city parks.

Red-bellied Woodpecker. *Melanerpes carolinus.* Southeastern South Dakota and forested valleys of Nebraska, Kansas, and Oklahoma to eastern Texas. Open deciduous woods, resident. Increasing nationally.

Yellow-bellied Sapsucker. *Sphyrapicus varius.* Eastern North Dakota and Black Hills. Deciduous and mixed woods, largely resident.

Red-naped Sapsucker. *Sphyrapicus nuchalis.* Black Hills. Coniferous forest, largely resident.

Ladder-backed Woodpecker. *Picoides scalaris.* Southwestern Kansas south to Texas and western Oklahoma. Riparian woods and arid woodlands, resident. Declining nationally.

Downy Woodpecker. *Picoides pubescens.* Entire region. Deciduous and mixed woods, parks, resident.

Hairy Woodpecker. *Picoides villosus.* Entire region. Open forests, resident. Increasing nationally.

Red-cockaded Woodpecker. *Picoides borealis.* Southeastern Oklahoma (McCurtain Co.). Open pine woods, resident. Nationally endangered.

Three-toed Woodpecker. *Picoides tridactylus.* Black Hills, where rare. Burned coniferous forest, resident. Probably increasing nationally.

Black-backed Woodpecker. *Picoides arcticus.* Black Hills, where uncommon. Burned coniferous forest, resident.

Northern Flicker. *Colaptes auratus.* Entire region. Diverse habitats; winters

uncommonly in region. Yellow-shafted race intergrades westward with red-shafted. Declining nationally.
Pileated Woodpecker. *Dryocopus pileatus.* Eastern North Dakota, eastern Kansas, and eastern half of Oklahoma. Rare in South Dakota and Nebraska. Mature forest, resident. Increasing nationally.

Family Tyrannidae—Tyrant Flycatchers
Western Wood-pewee. *Contopus sordidulus.* Western fourth of entire region. Coniferous and mixed forests. Declining nationally.
Eastern Wood-pewee. *Contopus virens.* Eastern half of entire region. Deciduous and mixed forests. Probably hybridizes locally with *sordidulus.* Declining nationally.
Acadian Flycatcher. *Empidonax virescens.* Southeastern South Dakota south to eastern parts of Nebraska, Kansas, and Oklahoma. Mature riparian forest.
Willow Flycatcher. *Empidonax traillii.* North Dakota south to central Nebraska, eastern Kansas, and northern Oklahoma. Swampy willow thickets.
Least Flycatcher. *Empidonax minimus.* North Dakota south to northeastern Nebraska. Deciduous woods. Declining nationally.
Dusky Flycatcher. *Empidonax oberholseri.* Black Hills. Open coniferous forest.
Cordilleran Flycatcher. *Empidonax occidentalis.* Black Hills and northwestern Nebraska. Riparian woods and canyons.
Eastern Phoebe. *Sayornis phoebe.* Eastern half of entire region. Open woods. Increasing nationally.
Say's Phoebe. *Sayornis saya.* Western half of entire region. Arid areas with cliffs. Probably increasing nationally.
Ash-throated Flycatcher. *Myiarchus cinerascens.* Texas and southwestern Oklahoma. Arid scrub and woodlands.
Great Crested Flycatcher. *Myiarchus crinitus.* Eastern half of entire region. Deciduous woods,
Cassin's Kingbird. *Tyrannus vociferans.* Western South Dakota, western Nebraska, and extreme western Oklahoma. Dry grasslands and scrub.
Western Kingbird. *Tyrannus verticalis.* Entire region, rarer eastward. Open country with scattered trees.
Eastern Kingbird. *Tyrannus tyrannus.* Entire region. Woodland edges, city parks. Declining nationally.
Scissor-tailed Flycatcher. *Tyrannus forficatus.* Texas and Oklahoma north to northern Kansas, occasionally to Nebraska. Open country.

Family Laniidae—Shrikes
Loggerhead Shrike. *Lanius ludovicianus.* Entire region. Open country, scattered trees; winters uncommonly in region. Declining nationally and regionally;

Great Plains race *migrans* is a candidate for federal listing as threatened or endangered.

Family Vireonidae—Vireos

White-eyed Vireo. *Vireo griseus*. Southeastern Nebraska to southwestern Oklahoma. Moist thickets.

Bell's Vireo. *Vireo bellii*. Central North Dakota to Oklahoma and Texas. Brushy thickets, riparian scrub. Declining nationally and regionally.

Black-capped Vireo. *Vireo atricapillus*. Central Oklahoma, previously north to Kansas. Brushy scrub oak woodlands. Nationally endangered.

Yellow-throated Vireo. *Vireo flavifrons*. Eastern North Dakota to eastern Oklahoma. Mature woods. Increasing nationally.

Plumbeous Vireo. *Vireo plumbeus*. Black Hills and northwestern Nebraska. Mixed woods. Increasing nationally.

Warbling Vireo. *Vireo gilvus*. Entire region, mainly eastward. Mature woods. Increasing nationally.

Philadelphia Vireo. *Vireo philadelphicus*. Northeastern North Dakota. Aspen woods, riparian thickets. Increasing nationally.

Red-eyed Vireo. *Vireo olivaceus*. North Dakota to western Oklahoma, mostly eastward. Mature deciduous forests and riparian wood. Increasing nationally.

Family Corvidae—Jays, Magpies, and Crows

Gray Jay. *Perisoreus canadensis*. Black Hills. Coniferous and mixed forests, resident.

Blue Jay. *Cyanocitta cristata*. Entire region. Forests, parks, residential areas, resident. Declining nationally.

Western Scrub-jay. *Aphelocoma californica*. Western Oklahoma. Oak-juniper woodland, resident. Increasing nationally.

Pinyon Jay. *Gymnorhinus cyanocephalus*. Western South Dakota, northwestern Nebraska, and western Oklahoma. Pinyon-juniper woodland, resident. Declining nationally.

Clark's Nutcracker. *Nucifraga columbiana*. Occasional in Black Hills. Coniferous woods, resident.

Black-billed Magpie. *Pica pica*. Western North Dakota to central Nebraska and west-central Kansas. Open country, diverse habitats, resident. Declining nationally.

American Crow. *Corvus brachyrhynchos*. Entire region. Diverse habitats, winters commonly in region. Increasing nationally.

Fish Crow. *Corvus ossifragus*. Southeastern Oklahoma. Woods near water, resident.

242

Chihuahuan Raven. *Corvus cryptoleucus.* Southwestern Kansas south to Texas and western Oklahoma. Arid scrub, resident. Probably declining nationally.

Common Raven. *Corvus corax.* Extreme western Oklahoma. Cliffs in open country, resident. Increasing nationally.

Family Alaudidae—Larks

Horned Lark. *Eremophila alpestris.* Entire region. Shorter grasslands; winters commonly in region. Declining nationally and regionally.

Family Hirundinidae—Swallows

Purple Martin. *Progne subis.* Eastern half of entire region. Widespread near water.

Tree Swallow. *Tachycineta bicolor.* North Dakota south to eastern Kansas. Open areas near trees. Probably increasing regionally.

Violet-green Swallow. *Tachycineta thalassina.* Western South Dakota and northwestern Nebraska. Open areas near cliffs. Probably increasing nationally.

Northern Rough-winged Swallow. *Stelgidopteryx serripennis.* Entire region. Open areas, often near water or cliffs.

Bank Swallow. *Riparia riparia.* North Dakota south to extreme northern Oklahoma. Open areas with clay banks.

Cliff Swallow. *Petrochelidon pyrrhonota.* Entire region. Open areas, often near cliffs or bridges. Probably increasing nationally and regionally.

Barn Swallow. *Hirundo rustica.* Entire region. Open areas near buildings. Declining nationally.

Family Paridae—Titmice

Carolina Chickadee. *Poecile carolinensis.* Southern Kansas to Oklahoma and Texas. Deciduous woods, resident. Hybridizes locally with *atricapillus.* Declining nationally.

Black-capped Chickadee. *Poecile atricapillus.* North Dakota to central Kansas, mainly eastward. Deciduous and mixed woods, resident. Increasing nationally.

Juniper Titmouse. *Baeolophus griseus.* Extreme western Oklahoma. Pinyon-juniper woodlands, resident.

Tufted Titmouse. *Baeolophus bicolor.* Northeastern Nebraska to western Oklahoma, mainly eastward. Deciduous forests and woodlands, resident. Increasing nationally.

Verdin. *Auriparus flaviceps.* Extreme southwestern Oklahoma. Arid scrub, resident. Declining nationally.

Bushtit. *Psaltriparus minimus.* Western Oklahoma, Texas panhandle. Dry brush and woodlands, resident. Probably declining nationally.

Family Sittidae—Nuthatches
Red-breasted Nuthatch. *Sitta canadensis.* Western South Dakota; rare in western Kansas. Coniferous and mixed woods; winters commonly in region. Increasing nationally.
White-breasted Nuthatch. *Sitta carolinensis.* North Dakota to central Oklahoma, mainly eastward. Deciduous and mixed woods, resident. Increasing nationally.
Pygmy Nuthatch. *Sitta pygmaea.* Black Hills and northwestern Nebraska. Ponderosa pine woods, resident.
Brown-headed Nuthatch. *Sitta pusilla.* Extreme southeastern Oklahoma. Pine forests and woodlands, resident. Declining nationally.

Family Certhiidae—Creepers
Brown Creeper. *Certhia americana.* Black Hills; locally elsewhere. Forests, resident.

Family Troglodytidae—Wrens
Rock Wren. *Salpinctes obsoletus.* Western North Dakota to western Oklahoma and Texas. Arid scrub with rocky areas, largely resident. Declining nationally.
Canyon Wren. *Catherpes mexicanus.* Black Hills; also western Oklahoma and Texas. Arid scrub with cliffs and canyons, resident.
Carolina Wren. *Thryothorus ludovicianus.* Southeastern Nebraska to Oklahoma and Texas. Deciduous wood, parks, residential areas, resident. Increasing nationally.
Bewick's Wren. *Thryomanes bewickii.* Northern Kansas to Oklahoma and Texas. Open country, riparian thickets, resident. Probably declining regionally.
House Wren. *Troglodytes aedon.* North Dakota to central Oklahoma and Texas. Open woods, riparian edges, residential areas, resident. Increasing nationally.
Sedge Wren. *Cistothorus platensis.* North Dakota to eastern Kansas; rare in South Dakota. Wet meadows with sedges; winters uncommonly in region. Increasing nationally.
Marsh Wren. *Cistothorus palustris.* North Dakota to Nebraska. Rarely to Kansas. Marshes; winters uncommonly in region. Increasing nationally.

Family Cinclidae—Dippers
American Dipper. *Cinclus mexicanus.* Black Hills. Rapid, cold streams, resident.

Family Regulidae—Kinglets
Golden-crowned Kinglet. *Regulus satrapa.* Black Hills; very rare in North Dakota. Coniferous forests; winters commonly in region.

Ruby-crowned Kinglet. *Regulus calendula.* Black Hills. Coniferous and mixed forests; winters commonly in region. Declining nationally.

Family Sylviidae—Gnatcatchers
Blue-gray Gnatcatcher. *Polioptila caerulea.* Southeastern South Dakota to western Oklahoma and Texas, mainly eastward. Deciduous forests and wooded areas. Increasing nationally.

Family Turdidae—Thrushes and Allies
Eastern Bluebird. *Sialia sialis.* North Dakota to western Oklahoma and Texas, mainly eastward. Forest edges; winters uncommonly in region. Increasing nationally.

Mountain Bluebird. *Sialia currucoides.* Western North Dakota to western Nebraska. Open coniferous forest; winters uncommonly in region. Hybridizes rarely with *sialis.* Increasing nationally.

Townsend's Solitaire. *Myadestes townsendi.* Black Hills and northwestern Nebraska. Coniferous forests; winters near breeding areas.

Veery. *Catharus fuscescens.* Central and eastern North Dakota; also Black Hills. Moist woods. Declining nationally.

Swainson's Thrush. *Catharus ustulatus.* Black Hills. Coniferous forests. Declining nationally.

Wood Thrush. *Hylocichla mustelina.* Eastern North Dakota to central Oklahoma, mainly eastward. Mature deciduous forests. Declining nationally.

American Robin. *Turdus migratorius.* Entire region. Open woods, parks, residential areas; winters uncommonly in region. Increasing nationally.

Family Mimidae—Mockingbirds and Thrashers
Gray Catbird. *Dumetella carolinensis.* Entire region. Shrubs and thickets.

Northern Mockingbird. *Mimus polyglottos.* Oklahoma and Texas north to northern Nebraska, occasionally to the Dakotas. Thickets, edges, residential areas; winters uncommonly in region. Declining nationally.

Sage Thrasher. *Oreoscoptes montanus.* Extreme western South Dakota and northwestern Nebraska (very rare). Sage scrub.

Brown Thrasher. *Toxostoma rufum.* Entire region. Deciduous edges, parks, and gardens; winters uncommonly in region. Declining nationally.

Curve-billed Thrasher. *Toxostoma curvirostre.* Rare in western Kansas. Arid scrub. Declining nationally.

Family Sturnidae—Starlings
European Starling. *Sturnus vulgaris.* Entire region; introduced. Diverse habitats, resident. Declining nationally.

Family Motacillidae—Pipits

Sprague's Pipit. *Anthus spragueii.* North Dakota south to central South Dakota. Short-grass and mixed-grass prairies. Declining nationally.

Family Bombycillidae—Waxwings

Cedar Waxwing. *Bombycilla cedrorum.* North Dakota south to northeastern Kansas; occasionally or rarely to Oklahoma. Open woods, parks; winters uncommonly in region. Increasing nationally.

Family Parulidae—Wood Warblers

Blue-winged Warbler. *Vermivora pinus.* North Dakota to central Oklahoma. Brushy wooded areas.

Golden-winged Warbler. *Vermivora chrysoptera.* Northeastern North Dakota (rare). Brushy wooded areas. Declining nationally.

Orange-crowned Warbler. *Vermivora celata.* North Dakota. Deciduous and mixed woods. Declining nationally.

Virginia's Warbler. *Vermivora virginiae.* Black Hills. Arid canyons and thickets.

Northern Parula. *Parula americana.* Southeastern Nebraska to eastern Oklahoma. Swampy wooded areas. Increasing nationally.

Yellow Warbler. *Dendroica petechia.* Entire region. Riparian wooded areas. Increasing nationally.

Chestnut-sided Warbler. *Dendroica pensylvanica.* Northern North Dakota, rarely South Dakota. Brushy woods. Declining nationally.

Yellow-rumped Warbler. *Dendroica coronata.* Black Hills, also southwestern North Dakota and northwestern Nebraska. Diverse wooded areas. Increasing nationally; "Myrtle's" and "Audubon's" races intergrade regionally.

Yellow-throated Warbler. *Dendroica dominica.* Southeastern Kansas to eastern Oklahoma. Bottomland deciduous woods.

Prairie Warbler. *Dendroica discolor.* Eastern Kansas to central Oklahoma. Brushy woods. Declining nationally.

Cerulean Warbler. *Dendroica cerulea.* Southeastern Nebraska (possibly eastern South Dakota) to eastern Oklahoma. Moist mature woods. Declining nationally; candidate for federal listing as threatened or endangered.

Black-and-white Warbler. *Mniotilta varia.* North Dakota to central Oklahoma. Scrubby woods or edges.

American Redstart. *Setophaga ruticilla.* North Dakota to eastern Oklahoma. Mature deciduous or mixed woods.

Prothonotary Warbler. *Protonotaria citrea.* Extreme eastern Nebraska to central Oklahoma. Swampy areas. Probably declining nationally.

Worm-eating Warbler. *Helmitheros vermivorus.* Southeastern Oklahoma. Dense, damp woods.

Swainson's Warbler. *Limnothlypis swainsonii.* Eastern Oklahoma. Moist wooded areas. Probably increasing nationally.

Ovenbird. *Seiurus aurocapillus.* North Dakota to eastern Oklahoma. Mature deciduous forests. Increasing nationally.

Northern Waterthrush. *Seiurus noveboracensis.* Northern North Dakota. Thickets near water.

Louisiana Waterthrush. *Seiurus motacilla.* Eastern Kansas to central Oklahoma. Wooded streams. Increasing nationally.

Kentucky Warbler. *Oporornis formosus.* Southeastern Nebraska to central Oklahoma. Moist mature woods. Declining nationally.

Mourning Warbler. *Oporornis philadelphia.* Northern North Dakota. Brushy deciduous woods. Declining nationally.

MacGillivray's Warbler. *Oporornis tolmiei.* Black Hills. Thick woods and edges.

Common Yellowthroat. *Geothlypis trichas.* Entire region. Marshy edges. Declining nationally.

Hooded Warbler. *Wilsonia citrina.* Extreme eastern Kansas and eastern Oklahoma. Dense deciduous woods, often near water.

Yellow-breasted Chat. *Icteria virens.* North Dakota to Oklahoma and Texas. Scrubby woods and pastures.

Family Thraupidae—Tanagers

Summer Tanager. *Piranga rubra.* Southeastern Nebraska to central Oklahoma. Mature deciduous forests.

Scarlet Tanager. *Piranga olivacea.* North Dakota to eastern Oklahoma. Mature deciduous forests.

Western Tanager. *Piranga ludoviciana.* Western South Dakota and northwestern Nebraska. Coniferous and mixed woods.

Family Emberizidae—Towhees, Sparrows, and Longspurs

Spotted Towhee. *Pipilo maculatus.* Western North Dakota to western Kansas, west of *erythropthalmus* in similar habitats. Hybrids common in region; winters uncommonly in region.

Eastern Towhee. *Pipilo erythropthalmus.* North Dakota to northeastern Oklahoma, mostly eastward. Riparian woods and thickets; winters uncommonly in region. Declining nationally.

Cassin's Sparrow. *Aimophila cassinii.* Southwestern Kansas, western Oklahoma, and Texas. Arid grasslands with scattered scrub. Declining nationally and regionally.

Bachman's Sparrow. *Aimophila aestivalis.* Southeastern Oklahoma. Open pine woods. Declining nationally.

Rufous-crowned Sparrow. *Aimophila ruficeps.* Central and western Oklahoma; rare in Kansas. Rocky scrub.

Chipping Sparrow. *Spizella passerina.* North Dakota to eastern Kansas and eastern Oklahoma. Open woods, parks, residential areas; winters uncommonly in region.

Clay-colored Sparrow. *Spizella pallida.* North Dakota to northeastern South Dakota, rarely farther south. Shrubby grasslands. Declining nationally.

Brewer's Sparrow. *Spizella breweri.* Western North Dakota, South Dakota, Nebraska, and extreme western Kansas, Oklahoma, and Texas. Sage scrub. Declining nationally and regionally.

Field Sparrow. *Spizella pusilla.* Western and central North Dakota to western Oklahoma. Overgrown fields and pastures; winters uncommonly in region. Declining nationally.

Vesper Sparrow. *Pooecetes gramineus.* North Dakota to southern Nebraska and northeastern Kansas. Weedy pastures and fields; winters uncommonly in region. Declining nationally and regionally.

Lark Sparrow. *Chondestes grammacus.* Entire region. Grasslands with scattered shrubs. Declining nationally.

Lark Bunting. *Calamospiza melanocorys.* North Dakota to western Oklahoma and Texas. Short-grass and sagesteppe. Declining nationally and regionally.

Savannah Sparrow. *Passerculus sandwichensis.* North Dakota to northwestern Nebraska. Tallgrass and mixed-grass prairie; winters uncommonly in region. Declining nationally.

Baird's Sparrow. *Ammodramus bairdii.* North Dakota and northern South Dakota. Mixed-grass prairie. Declining nationally and regionally; candidate for federal listing as threatened or endangered.

Grasshopper Sparrow. *Ammodramus savannarum.* Entire region but rare in Texas. Mixed-grass prairie. Declining nationally and regionally.

Henslow's Sparrow. *Ammodramus henslowii.* Southeastern South Dakota to eastern Kansas. Weedy fields and pastures. Declining nationally; candidate for federal listing as threatened or endangered.

Le Conte's Sparrow. *Ammodramus leconteii.* Eastern North Dakota. Sedge meadows. Probably increasing nationally.

Nelson's Sharp-tailed Sparrow. *Ammodramus nelsoni.* Eastern and northern North Dakota; rare in South Dakota. Marshes and wet meadows. Probably increasing nationally.

Song Sparrow. *Melospiza melodia.* North Dakota to eastern Nebraska, locally farther south to central Kansas. Brushy fields, riparian edges; winters commonly in region. Declining nationally.

Swamp Sparrow. *Melospiza georgiana.* Eastern Dakotas, local in Nebraska. Marshes, fens, wet meadows; winters uncommonly in region. Increasing nationally.

White-throated Sparrow. *Zonotrichia albicollis.* Northeastern North Dakota.

Coniferous and mixed woods; winters commonly in region. Declining nationally.

Dark-eyed Junco. *Junco hyemalis.* Black Hills, adjacent South Dakota, and Nebraska. Ponderosa pine woods; winters commonly in region. Black Hills ("white-winged") race *aikeni* endemic to region. Declining nationally.

McCown's Longspur. *Calcarius mccownii.* Western North Dakota and extreme western Nebraska; one South Dakota record. Short-grass prairie. Probably declining nationally.

Chestnut-collared Longspur. *Calcarius ornatus.* North Dakota south to northwestern Nebraska. Mixed-grass prairie. Declining nationally, probably also regionally.

Family Cardinalidae—Cardinals, Grosbeaks, and Allies

Northern Cardinal. *Cardinalis cardinalis.* Southeastern North Dakota to Texas and western Oklahoma. Cities and open forests, resident.

Rose-breasted Grosbeak. *Pheucticus ludovicianus.* Eastern North Dakota to central Oklahoma. Open deciduous forests. Declining nationally.

Black-headed Grosbeak. *Pheucticus melanocephalus.* Western North Dakota to southwestern Kansas. Open forests and woodlands west of *ludovicianus,* sometimes hybridizing.

Blue Grosbeak. *Guiraca caerulea.* Central South Dakota (rarely North Dakota) to Texas and Oklahoma. Scrubby woods, weedy fields. Increasing nationally.

Lazuli Bunting. *Passerina amoena.* Western North Dakota to Texas and western Oklahoma. Drier woods and brushy edges.

Indigo Bunting. *Passerina cyanea.* North Dakota to central Oklahoma. Deciduous woods and edges east of *amoena,* sometimes hybridizing. Declining nationally.

Painted Bunting. *Passerina ciris.* Southeastern Kansas to western Oklahoma. Riparian thickets, weedy areas. Declining nationally.

Dickcissel. *Spiza americana.* Entire region. Tallgrass and mixed-grass prairie. Declining nationally and regionally.

Family Icteridae—Meadowlarks, Blackbirds, Orioles, and Allies

Bobolink. *Dolichonyx oryzivorus.* North Dakota to southern Nebraska, local in Kansas. Wet meadows and low prairies. Declining nationally.

Red-winged Blackbird. *Agelaius phoeniceus.* Entire region. Diverse habitats; winters commonly in region. Declining nationally.

Brewer's Blackbird. *Euphagus cyanocephalus.* North Dakota to western Nebraska. Diverse habitats. Declining nationally.

Eastern Meadowlark. *Sturnella magna.* Southeastern South Dakota to western

Oklahoma, locally to western Nebraska and Texas. Tallgrass prairies; winters uncommonly in region. Declining nationally and regionally.

Western Meadowlark. *Sturnella neglecta.* Entire region except eastern Kansas and eastern Oklahoma, where occasional. Mixed-grass and short-grass prairies, rarely hybridizing with *magna;* winters uncommonly in region. Declining nationally.

Yellow-headed Blackbird. *Xanthocephalus xanthocephalus.* North Dakota south to southern Kansas, occasionally to Texas. Larger, deeper marshes. Increasing nationally.

Common Grackle. *Quiscalus quiscula.* Entire region. Woody edges, parks, residential areas; winters uncommonly in region. Declining nationally.

Great-tailed Grackle. *Quiscalus mexicanus.* Texas and Oklahoma north to central Nebraska, rarely to South Dakota; expanding northward. Parks, residential areas; winters uncommonly in region. Increasing nationally.

Brown-headed Cowbird. *Molothrus ater.* Entire region. Open habitats and edges; winters uncommonly in region. Declining nationally.

Orchard Oriole. *Icterus spurius.* Almost entire region, occasional in Texas. Orchards, second-growth.

Baltimore Oriole. *Icterus galbula.* North Dakota south to western Oklahoma and eastern Texas. Riparian woods, forest edges, parks. Declining nationally.

Bullock's Oriole. *Icterus bullockii.* Southwestern North Dakota south to Texas and extreme western Oklahoma. West of *galbula* in similar habitats, sometimes hybridizing. Declining nationally.

Family Fringillidae—Finches
House Finch. *Carpodacus mexicanus.* Entire region. Residential areas, resident. Probably increasing nationally; Great Plains range still expanding from east.

Cassin's Finch. *Carpodacus cassini.* Black Hills. Dry pine forests, resident. Declining nationally.

Red Crossbill. *Loxia curvirostra.* Black Hills; occasionally elsewhere in the Dakotas and south erratically to Kansas. Pine forests and woodlands, largely resident.

Pine Siskin. *Carduelis pinus.* Black Hills, western Nebraska, and occasionally elsewhere, south at least to Kansas. Residential areas, coniferous and mixed woods; winters commonly in region.

Lesser Goldfinch. *Carduelis psaltria.* Extreme western Oklahoma; rare in Kansas. Open areas with scattered brush or trees, resident. Probably declining nationally.

American Goldfinch. *Carduelis tristis.* Entire region except for southern Oklahoma and Texas, where rare. Weedy fields. Largely resident. Declining nationally.

Evening Grosbeak. *Coccothraustes vespertinus.* Black Hills, where rare. Coniferous and mixed woods, largely resident.

Family Passeridae—Old World Sparrows
House Sparrow. *Passer domesticus.* Entire region, introduced. Habitats near
   humans, resident. Declining nationally.

Transient Bird Species
The following species rarely if ever breed in the plains states, but most are
annual spring and fall transients ("regular migrants"). A few are overwintering migrants from breeding areas farther north, and others are "vagrants" out
of their normal ranges with only irregular occurrences in the region. Some are
nonbreeding summer "visitors."

Family Ardeidae—Bitterns and Herons
Tricolored Heron. *Egretta tricolor.* Occasional vagrant southward; very rare or
   accidental breeder throughout.

Family Ciconiidae—Storks
Wood Stork. *Mycteria americana.* Occasional vagrant southward; endangered
   nationally.

Family Anatidae—Swans, Geese, and Ducks
Black-bellied Whistling-duck. *Dendrocygna autumnalis.* Occasional vagrant
   southward.
Fulvous Whistling-duck. *Dendrocygna bicolor.* Occasional vagrant southward.
Greater White-fronted Goose. *Anser albifrons.* Regular migrant; winters uncom-
   monly in region.
Snow Goose. *Chen caerulescens.* Regular migrant; winters uncommonly in region.
Ross's Goose. *Chen rossii.* Regular migrant; winters uncommonly in region.
Tundra Swan. *Cygnus columbianus.* Regular migrant northward.
Eurasian Wigeon. *Anas penelope.* Casual vagrant.
Surf Scoter. *Melanitta perspicillata.* Occasional vagrant.
White-winged Scoter. *Melanitta fusca.* Regular migrant northward.
Black Scoter. *Melanitta nigra.* Casual vagrant.
Long-tailed Duck (Oldsquaw). *Clangula hyemalis.* Regular overwintering
   migrant.
Barrow's Goldeneye. *Bucephala islandica.* Casual migrant westward.
Red-breasted Merganser. *Mergus serrator.* Regular migrant.

Family Accipitridae—Kites, Hawks, Eagles, and Allies
White-tailed Kite. *Elanus leucurus.* Casual vagrant southward.
Harris's Hawk. *Parabuteo unicinctus.* Occasional vagrant southward.
Rough-legged Hawk. *Buteo lagopus.* Regular overwintering migrant northward.

Family Falconidae—Falcons
Gyrfalcon. *Falco rusticolus.* Occasional overwintering migrant northward.

Family Charadriidae—Plovers
Black-bellied Plover. *Pluvialis squatarola*. Regular migrant.
American Golden-plover. *Pluvialis dominica*. Regular migrant.
Semipalmated Plover. *Charadrius semipalmatus*. Regular migrant.

Family Scolopacidae—Sandpipers and Phalaropes
Greater Yellowlegs. *Tringa melanoleuca*. Regular migrant.
Lesser Yellowlegs. *Tringa flavipes*. Regular migrant.
Solitary Sandpiper. *Tringa solitaria*. Regular migrant.
Whimbrel. *Numenius phaeopus*. Occasional spring migrant.
Hudsonian Godwit. *Limosa haemastica*. Regular spring migrant, rare in fall.
Ruddy Turnstone. *Arenaria interpres*. Regular spring migrant, rare in fall.
Red Knot. *Calidris canutus*. Occasional migrant.
Sanderling. *Calidris alba*. Regular migrant.
Semipalmated Sandpiper. *Calidris pusilla*. Regular migrant.
Western Sandpiper. *Calidris mauri*. Regular migrant.
Least Sandpiper. *Calidris minutilla*. Regular migrant.
White-rumped Sandpiper. *Calidris fuscicollis*. Regular spring migrant, rare in fall.
Baird's Sandpiper. *Calidris bairdii*. Regular migrant.
Pectoral Sandpiper. *Calidris melanotos*. Regular migrant.
Dunlin. *Calidris alpina*. Regular spring migrant, rare in fall.
Stilt Sandpiper. *Calidris himantopus*. Regular migrant.
Buff-breasted Sandpiper. *Tryngites subruficollis*. Regular migrant.
Short-billed Dowitcher. *Limnodromus griseus*. Occasional migrant.
Long-billed Dowitcher. *Limnodromus scolopaceus*. Regular migrant.
Red-necked Phalarope. *Phalaropus lobatus*. Regular migrant.
Red Phalarope. *Phalaropus fulicaria*. Occasional migrant.

Family Laridae—Gulls and Terns
Laughing Gull. *Larus atricilla*. Casual vagrant southward.
Bonaparte's Gull. *Larus philadelphia*. Regular migrant.
Herring Gull. *Larus argentatus*. Regular migrant and summer visitor; winters
  uncommonly in region.
Glaucous Gull. *Larus hyperboreus*. Occasional vagrant.
Caspian Tern. *Sterna caspia*. Regular migrant and summer visitor; possible rare
  breeder (South Dakota).

Family Strigidae—Typical Owls
Snowy Owl. *Nyctea scandiaca*. Regular overwintering migrant northward.
Northern Hawk-owl. *Surnia ulula*. Casual overwintering vagrant northward.
Great Gray Owl. *Strix nebulosa*. Casual overwintering vagrant northward.
Boreal Owl. *Aegolius funereus*. Casual overwintering vagrant northward.

Family Trochilidae—Hummingbirds
Calliope Hummingbird. *Stellula calliope.* Occasional migrant westward.
Rufous Hummingbird. *Selasphorus rufus.* Occasional migrant westward.

Family Tyrannidae—Tyrant Flycatchers
Olive-sided Flycatcher. *Contopus cooperi.* Regular migrant; possible rare breeder in Black Hills.
Yellow-bellied Flycatcher. *Empidonax flaviventris.* Regular migrant.
Alder Flycatcher. *Empidonax alnorum.* Regular migrant.
Hammond's Flycatcher. *Empidonax hammondii.* Occasional migrant westward.
Vermilion Flycatcher. *Pyrocephalus rubinus.* Occasional vagrant southward.

Family Laniidae—Shrikes
Northern Shrike. *Lanius excubitor.* Regular overwintering migrant northward.

Family Vireonidae—Vireos
Blue-headed Vireo. *Vireo solitarius.* Regular migrant eastward.

Family Troglodytidae—Wrens
Winter Wren. *Troglodytes troglodytes.* Regular migrant; possible breeder in Black Hills.

Family Turdidae—Thrushes and Allies
Gray-cheeked Thrush. *Catharus minimus.* Regular migrant.
Hermit Thrush. *Catharus guttatus.* Regular migrant; very rare breeder in Black Hills.
Varied Thrush. *Ixoreus naevius.* Occasional migrant westward.

Family Motacillidae—Pipits
American Pipit. *Anthus rubescens.* Regular migrant.

Family Bombycillidae—Waxwings
Bohemian Waxwing. *Bombycilla garrulus.* Occasional overwintering migrant northward.

Family Parulidae—Wood Warblers
Golden-winged Warbler. *Vermivora chrysoptera.* Regular migrant.
Tennessee Warbler. *Vermivora peregrina.* Regular migrant.
Nashville Warbler. *Vermivora ruficapilla.* Regular migrant.
Yellow Warbler. *Dendroica petechia.* Regular migrant.
Magnolia Warbler. *Dendroica magnolia.* Regular migrant.
Cape May Warbler. *Dendroica tigrina.* Regular migrant.
Black-throated Blue Warbler. *Dendroica caerulescens.* Occasional migrant eastward.
Townsend's Warbler. *Dendroica townsendi.* Occasional vagrant westward.

Black-throated Green Warbler. *Dendroica virens*. Regular migrant.
Blackburnian Warbler. *Dendroica fusca*. Regular migrant.
Pine Warbler. *Dendroica pinus*. Regular migrant.
Palm Warbler. *Dendroica palmarum*. Regular migrant.
Bay-breasted Warbler. *Dendroica castanea*. Regular migrant.
Blackpoll Warbler. *Dendroica striata*. Regular migrant.
Connecticut Warbler. *Oporornis agilis*. Regular migrant.
Wilson's Warbler. *Wilsonia pusilla*. Regular migrant.
Canada Warbler. *Wilsonia canadensis*. Regular migrant.

Family Emberizidae—Towhees, Sparrows, and Longspurs
Green-tailed Towhee. *Pipilo chlorurus*. Casual vagrant westward.
American Tree Sparrow. *Spizella arborea*. Regular overwintering migrant.
Black-throated Sparrow. *Amphispiza bilineata*. Occasional vagrant southward.
Sage Sparrow. *Amphispiza belli*. Occasional migrant westward.
Fox Sparrow. *Passerella iliaca*. Regular migrant.
Lincoln's Sparrow. *Melospiza lincolnii*. Regular migrant.
Harris's Sparrow. *Zonotrichia querula*. Regular overwintering migrant.
White-crowned Sparrow. *Zonotrichia leucophrys*. Regular overwintering migrant.
Golden-crowned Sparrow. *Zonotrichia atricapilla*. Occasional vagrant westward.
Lapland Longspur. *Calcarius lapponicus*. Regular overwintering migrant northward.
Smith's Longspur. *Calcarius pictus*. Regular overwintering migrant.
Snow Bunting. *Plectrophenax nivalis*. Regular overwintering migrant northward.

Family Icteridae—Meadowlarks, Blackbirds, Orioles, and Allies
Rusty Blackbird. *Euphagus carolinus*. Regular migrant.
Scott's Oriole. *Icterus parisorum*. Occasional vagrant southward.

Family Fringillidae—Finches
Gray-crowned Rosy-finch. *Leucosticte tephrocotis*. Occasional overwintering migrant westward.
Pine Grosbeak. *Pinicola enucleator*. Occasional overwintering migrant northward.
Purple Finch. *Carpodacus purpureus*. Regular migrant.
White-winged Crossbill. *Loxia leucoptera*. Occasional overwintering migrant northward; possible rare breeder in Black Hills.
Common Redpoll. *Carduelis flammea*. Regular overwintering migrant northward.
Hoary Redpoll. *Carduelis hornemanni*. Occasional overwintering migrant northward.

Mammals of the Great Plains States

This list of 140 species is based largely on Jones, Armstrong, and Choate (1985), whose mammal survey extended from North Dakota through Oklahoma. Texas comments here refer only to the panhandle region. Habitat descriptions should be considered minimal. Sixteen species endemic to or especially characteristic of the Great Plains (Jones et al. 1983) are underlined; about 80 species are regular residents of the Great Plains grassland ecosystem. This list excludes some introduced species, and some that are very rare or of uncertain current occurrence. The most recent taxonomic nomenclature may be found in Wilson and Ruff (1999) and in Whittaker (1996). Familial and generic sequences are arranged in the sequence used by Wilson and Ruff; species are listed alphabetically within genera. Some recent taxonomic changes are incorporated here and are explained or indicated parenthetically.

Important regional or state mammal surveys are available for Texas (Davis and Schmidly 1995; Choate 1997), Oklahoma (Caire et al. 1989), Kansas (Cockrum 1952; Bee et al. 1981), Nebraska (Jones 1964), and South Dakota (Turner 1974; Higgins et al. 2000). Bailey's (1928) survey of North Dakota is outdated. Surveys for adjoining states include Montana (Hoffman and Pattie 1968; Forsman 2001), Wyoming (Clark and Stromberg 1987), Colorado (Fitzgerald, Meanley, and Armstrong 1994), New Mexico (Findley 1987; Findley et al. 1975), Minnesota (Hazard 1982), Iowa (Bowles 1975), Missouri (Schwartz and Schwartz 1981), and Arkansas (Sealander 1979; Choate, Jones, and Jones 1994).

Family Didelphidae—New World Opossums
Virginia Opossum. *Didelphis virginiana*. Southeastern North Dakota to Oklahoma and Texas. Woods, farmlands.

Family Dasypodidae—Armadillos

Nine-banded Armadillo. *Dasypus novemcinctus*. Southern Nebraska (rare) to Oklahoma and Texas. Diverse habitats.

Family Soricidae—Shrews

Arctic Shrew. *Sorex arcticus*. Northwestern North Dakota to northeastern South Dakota (no recent records). Marshes, grassy clearings.

Masked Shrew. *Sorex cinereus*. North Dakota to Nebraska and extreme northern Kansas. Generally moist habitats. The form *haydeni*, or Hayden's shrew, has at times been separated from *cinereus*, and their relative ranges are still uncertain.

Pygmy Shrew. *Sorex hoyi*. Eastern North Dakota and eastern South Dakota. Diverse habitats.

Merriam's Shrew. *Sorex merriami*. Rare in North Dakota and Nebraska; a few South Dakota records. Arid habitats.

Dwarf Shrew. *Sorex nanus*. Western South Dakota (extremely rare). Diverse habitats, especially grasslands.

Water shrew. *Sorex palustris*. Eastern North Dakota to northeastern South Dakota. Aquatic habitats.

Northern Short-tailed Shrew. *Blarina brevicauda*. Eastern North Dakota to eastern Nebraska. Woods and grasslands.

Southern Short-tailed Shrew. *Blarina carolinensis*. Southern Nebraska to Oklahoma. Moist hardwoods.

Elliot's Short-tailed Shrew. *Blarina hylophaga*. Southeastern Oklahoma. Tall grasses.

Least Shrew. *Cryptotis parva*. Central and western South Dakota to Oklahoma and Texas. Weedy or brushy fields.

Desert Shrew. *Notiosorex crawfordi*. Extreme southwestern Kansas, Oklahoma, and Texas. Mostly arid habitats.

Family Talpidae—Moles

Eastern Mole. *Scalopus aquaticus*. Southern South Dakota to Oklahoma and Texas. Loamy soils.

Family Vespertilionidae—Vespertilionid Bats

Southeastern Myotis. *Myotis austroriparius*. Southeastern Oklahoma. Near water.

Western Small-footed Myotis. *Myotis ciliolabrum* (previously also included *leibii*, which is now considered a separate species). Western North Dakota to western Oklahoma and Texas. Rocky outcrops.

Long-eared Myotis. *Myotis evotis*. Western North Dakota and northwestern South Dakota. Mostly coniferous woods.

Gray Myotis. *Myotis grisescens.* Eastern Oklahoma and extreme southeastern Kansas (state endangered). Areas with limestone caverns.

Little Brown Myotis. *Myotis lucifugus.* North Dakota through eastern Nebraska and Kansas to eastern Oklahoma. Buildings, caves, mines.

Northern Myotis. *Myotis septentrionalis.* North Dakota and western South Dakota to eastern Oklahoma; also the Black Hills. Until 1979 considered a subspecies of *Myotis keeni,* now restricted to the Pacific Northwest. Buildings, caves, mines.

Indiana (Social) Myotis. *Myotis sodalis.* Eastern Oklahoma. Limestone caves.

Fringed Myotis. *Myotis thysanodes.* Southwestern South Dakota and northwestern Nebraska. Buildings, mines, caves. Very rare.

Cave Myotis. *Myotis velifer.* Western Kansas, western Oklahoma, and Texas. Caves.

Long-legged Myotis. *Myotis volans.* Western North Dakota, western South Dakota, and extreme northwestern Nebraska. Cliffs, pine forests.

Yuma Myotis. *Myotis yumanensis.* Extreme northwestern Oklahoma. Near water.

Silver-haired Bat. *Lasionycteris noctivagans.* Entire region. Forests, grasslands. Migratory.

Western Pipistrelle. *Pipistrellus hesperus.* Southwestern Oklahoma and Texas. Arid habitats.

Eastern Pipistrelle. *Pipistrellus subflavus.* Southeastern Nebraska to Oklahoma and Texas. Buildings, caves.

Big Brown Bat. *Eptesicus fuscus.* Entire region. Towns, farms, other buildings.

Eastern Red Bat. *Lasiurus borealis.* Entire region. Prairies, farms, woods. Migratory but hibernating locally.

Seminole Bar. *Lasiurus seminolus.* Extreme southeastern Oklahoma (rare).

Hoary Bat. *Lasiurus cinereus.* Entire region. Diverse habitats. Migratory.

Evening Bat. *Nycticeius humeralis.* Eastern Nebraska to eastern Oklahoma.

Rafinesque's Big-eared Bat. *Corynorhinus (Plecotus) rafinesquii.* Southeastern Oklahoma. Trees, crevices, caves.

Townsend's Big-eared Bat. *Corynorhinus (Plecotus) townsendii.* Black Hills and western North Dakota to western Nebraska, and southern Kansas to Oklahoma and Texas. Buildings, gypsum caves, caverns. Declining; race *ingens* endangered in Oklahoma.

Pallid Bat. *Antrozous pallidus.* Western Oklahoma and Texas. Arid habitats.

Family Molossidae—Molossid Bats

Brazilian Free-tailed Bat. *Tadarida brasiliensis.* Southeastern South Dakota and most of Nebraska to Oklahoma and Texas. Caverns, caves, buildings. Migratory.

Big Free-tailed Bat. *Tadarida (Nyctinomops) macrotis.* Scattered records; Kansas and south. Caves in arid country. Migratory.

Family Canidae—Coyotes, Wolves, and Foxes

Coyote. *Canis latrans.* Entire region. Diverse habitats.

Gray (Prairie) Wolf. *Canis lupus.* Historically extirpated but now occasional in northern North Dakota.

Red Wolf. *Canis rufus.* Extirpated from southern plains (Oklahoma southward).

Swift Fox. *Vulpes v. velox.* North Dakota to western Oklahoma and Texas. Rare and declining species. Few North Dakota records since 1970, very rare in South Dakota (critically imperiled), and state endangered (imperiled) in Nebraska. Short-grass plains. The kit fox (*V. m. macrotis*) is now considered conspecific with the swift fox.

Red Fox. *Vulpes vulpes.* Entire region. Woodland edges, farmlands.

Gray Fox. *Urocyon cineroargenteus.* Eastern North Dakota to western Oklahoma. Hardwood forests.

Family Mustelidae—Weasels, Badgers, Skunks, and Otters

Marten. *Martes americana.* Northeastern North Dakota (rare). Coniferous forests.

Fisher. *Martes pennanti.* Northeastern North Dakota (rare). Coniferous forests.

Ermine. *Mustela erminea.* Eastern North Dakota; also Black Hills. Diverse habitats.

Long-tailed Weasel. *Mustela frenata.* Entire region. Diverse habitats.

Black-footed Ferret. *Mustela nigriceps.* Reintroduced in southwestern South Dakota. Short-grass plains. Nearly extirpated; nationally endangered.

Least Weasel. *Mustela nivalis.* North Dakota to northern Kansas. Grassy fields.

Mink. *Mustela vison.* Entire region. Near water.

Wolverine. *Gulo gulo.* Extirpated; previously from North Dakota to Nebraska.

Badger. *Taxidea taxus.* Entire region except southeastern Oklahoma. Open country.

Northern River Otter. *Lontra canadensis.* Extirpated but locally reintroduced. Near water. State endangered in Nebraska, threatened in South Dakota.

Family Mephitidae—Skunks

Western Spotted Skunk. *Spilogale gracilis.* Extreme western Oklahoma. Open country and forest edges.

Eastern Spotted Skunk. *Spilogale putorius.* Entire region except northern North Dakota. Forests and forest edges. Declining species, threatened in Kansas.

Striped Skunk. *Mephitis mephitis.* Entire region. Forest edges.

Family Procyonidae—Raccoons and Allies

Raccoon. *Procyon lotor.* Entire region. Woods, suburbs.

Ringtail. *Bassariscus astutus.* Southern Kansas, Oklahoma, and Texas. Diverse habitats.

Family Felidae—Cats

Mountain Lion. *Puma concolor.* Very rare throughout region but probably increasing slowly. Open country.

Lynx. *Felis lynx.* North Dakota to (rarely) South Dakota and (very rarely) Nebraska. Coniferous woods.

Bobcat. *Felis rufus.* Entire region. Open, often rocky country.

Family Cervidae—Deer, Elk, and Moose

Elk (Wapiti). *Cervus elephas.* Historically extirpated, locally reintroduced in both confined and free-living locations. Large grasslands.

Mule Deer. *Odocoileus hemionus.* North Dakota to central Kansas, western Oklahoma, and Texas. Steppe, sagesteppe, and brushlands.

White-tailed Deer. *Odocoileus virginianus.* Entire region. Forest edges.

Moose. *Alces alces.* Northeastern North Dakota (rare). Boreal wetlands.

Family Antilocapridae—Pronghorns

Pronghorn. *Antilocapra americana.* Western North Dakota to western Oklahoma and Texas. Short-grass plains; reestablishing after near extirpation.

Family Bovidae—Cattle, Sheep, and Goats

Bison. *Bison (Bos) bison.* Historically extirpated but reintroduced widely in confined locations. Extensive grasslands.

Bighorn (Mountain) Sheep. *Ovis canadensis.* Endemic historic race now extinct, but species reintroduced in both Dakotas and Nebraska. Rimrock and montane areas.

Mountain Goat. *Oreamnos americanus.* Introduced (1924) and now resident in the Black Hills. Steep cliffs.

Family Sciuridae—Squirrels

Least Chipmunk. *Tamias minimus.* North Dakota, western South Dakota, and northwestern Nebraska. Pine woods, rocky areas.

Colorado Chipmunk. *Tamias quadrivittatus.* Extreme western Oklahoma. Pinyon-juniper woodlands.

Eastern Chipmunk. *Tamias striatus.* Eastern North Dakota to eastern Oklahoma. Deciduous woods.

Yellow-bellied Marmot. *Marmota flaviventris.* Black Hills of South Dakota. Rocky meadows.

Woodchuck. *Marmota monax.* Eastern North Dakota to eastern Oklahoma. Forest edges.

Wyoming Ground Squirrel. *Spermophilus elegans* (previously part of *S. richardsonii*). Western Nebraska. Sagesteppes.

Franklin's Ground Squirrel. *Spermophilus franklinii.* Eastern North Dakota to eastern Kansas. Tall grass and woody edges.

Richardson's Ground Squirrel. *Spermophilus richardsonii.* North Dakota to eastern South Dakota. Native shorter prairies.

Spotted Ground Squirrel. *Spermophilus spilosoma.* Southwestern South Dakota to western Oklahoma and Texas. Arid, sandy grasslands.

Thirteen-lined Ground Squirrel. *Spermophilus tridecemlineatus.* Entire region. Shorter grasslands and cut or grazed areas of taller grasslands.

Rock Squirrel. *Spermophilus variegatus.* Extreme western Oklahoma. Rocky habitats.

Black-tailed Prairie Dog. *Cynomys ludovicianus.* Western North Dakota to western Oklahoma and Texas. Short-grass and mixed-grass prairies. Nationally declining; locally threatened or endangered.

Eastern Gray Squirrel. *Sciurus carolinensis.* North Dakota to eastern Oklahoma, except South Dakota, where very rare in northeast. Mostly hardwood forests.

Eastern Fox Squirrel. *Sciurus niger.* Entire region. Mostly hardwood forests.

Red (Pine) Squirrel. *Tamiasciurus hudsonicus.* Eastern North Dakota and Black Hills. Coniferous forests.

Northern Flying Squirrel. *Glaucomys sabrinus.* Eastern North Dakota and Black Hills. Coniferous or mixed forests.

Southern Flying Squirrel. *Glaucomys volans.* Southeastern Nebraska (state threatened) to eastern Oklahoma. Deciduous or mixed forests.

Family Geomyidae—Pocket Gophers
Northern Pocket Gopher. *Thomomys talpoides.* North Dakota to western Nebraska. Diverse habitats with deep soils.

Baird's Pocket Gopher. *Geomys breviceps.* Eastern Oklahoma. Sandy soils.

Plains Pocket Gopher. *Geomys bursarius.* Eastern North Dakota to Oklahoma and Texas. Grasslands.

Yellow-faced Pocket Gopher. *Pappogeomys (Cratogeomys) castanops.* Western Kansas to western Oklahoma and Texas. Sandy soils.

Family Heteromyidae – Pocket Mice, Kangaroo Rats, and Kangaroo Mice
Olive-backed (Wyoming) Pocket Mouse. *Perognathus fasciatus.* North Dakota to western Nebraska. Steppes and sagesteppes.

Plains Pocket Mouse. *Perognathus flavescens.* Eastern North Dakota to western Oklahoma and Texas. Sandy areas.

Silky Pocket Mouse. *Perognathus flavus.* Extreme southwestern South Dakota to western Oklahoma and Texas. Semiarid grasslands.

Hispid Pocket Mouse. *Perognathus (Chaetodipus) hispidus.* Southwestern North-Dakota to Oklahoma and Texas. Sandy grasslands.

Texas Kangaroo Rat. *Dipodomys elator.* Southwestern Oklahoma, where possibly extirpated.

Ord's Kangaroo Rat. *Dipodomys ordii.* Western North Dakota to western Oklahoma and Texas. Sandy grasslands.

Family Castoridae—Beavers
Beaver. *Castor canadensis.* Entire region. Woods near water.

Family Muridae—Rats and Mice
Marsh Rice Rat. *Oryzomys palustris.* Southeastern Oklahoma. Marshes.

Fulvous Harvest Mouse. *Reithrodontomys fulvescens.* Southeastern Kansas and Oklahoma. Grassy fields with shrubs.

Eastern Harvest Mouse. *Reithrodontomys humulus.* Southeastern Oklahoma. Weedy fields.

Western Harvest Mouse. *Reithrodontomys megalotis.* North Dakota to western Oklahoma and Texas. Weedy fields and tallgrass prairies.

Plains Harvest Mouse. *Reithrodontomys montanus.* Western South Dakota to Oklahoma and Texas. Grassy habitats.

Texas Mouse. *Peromyscus attwateri.* Southeastern Kansas and Oklahoma. Rocky habitats.

Brush Mouse. *Peromyscus boylii.* Extreme western Oklahoma. Rocky areas.

Northern Rock Mouse. *Peromyscus (difficilis) nasutus.* Extreme western Oklahoma. Rocky areas.

Cotton Mouse. *Peromyscus gossypinus.* Southeastern Oklahoma. Moist habitats.

White-footed Mouse. *Peromyscus leucopus.* Entire region. Wooded habitats.

Deer Mouse. *Peromyscus maniculatus.* Entire region. Open habitats.

White-ankled Mouse. *Peromyscus pectoralis.* Extreme southern Oklahoma. Rocky areas.

Pinyon Mouse. *Peromyscus trueii.* Extreme western Oklahoma. Pinyon-juniper.

Golden Mouse. *Ochrotomys nuttallii.* Southeastern Oklahoma. Dense forests.

Northern Grasshopper Mouse. *Onychomys leucogaster.* North Dakota to western Oklahoma and Texas. Open grasslands.

Hispid Cotton Rat. *Sigmodon hispidus.* Southern Nebraska to Oklahoma and Texas. Grasslands and old fields.

White-throated Woodrat. *Neotoma albigula.* Extreme western Oklahoma. Arid habitats.

Bushy-tailed Woodrat. *Neotoma cinerea.* Western North Dakota to western Nebraska. Rocky areas.

Eastern Woodrat. *Neotoma floridana.* Niobrara Valley and southern Nebraska to eastern Oklahoma. Wooded areas.

Mexican Woodrat. *Neotoma mexicana.* Extreme western Oklahoma. Rocky areas.

Southern Plains Woodrat. *Neotoma micropus.* Southwestern Kansas to western Oklahoma and Texas. Diverse habitats.

Southern Red-backed Vole. *Clethrionomys gapperi.* North Dakota to northeastern South Dakota; also the Black Hills. Mesic habitats.

Long-tailed Vole. *Microtus longicaudus.* Black Hills of South Dakota. Diverse habitats.

Prairie Vole. *Microtus ochrogaster.* North Dakota to central Oklahoma. Grasslands.

Meadow Vole. *Microtus pennsylvanicus.* North Dakota to extreme northern Kansas. Grassy fields.

Woodland Vole. *Microtus pinetorum.* Southeastern Nebraska to Oklahoma. Deciduous woods.

Sagebrush Vole. *Lemniscus curtatus.* Western North Dakota and extreme northwestern South Dakota (not reported since 1971). Sagesteppes.

Muskrat. *Ondatra zibethicus.* Entire region. Marshes.

Southern Bog Lemming. *Synaptomys cooperi.* Southeastern South Dakota to south-central Kansas. Diverse habitats.

House Mouse. *Mus musculus.* Introduced; entire region. Near humans.

Norway Rat. *Rattus norvegicus.* Introduced; entire region. Near humans.

Black Rat. *Rattus rattus.* Introduced; widespread north to Kansas. Near humans.

Family Dipodidae—Jumping Mice

Meadow Jumping Mouse. *Zapus hudsonicus.* North Dakota to northeastern Oklahoma. Grassy or weedy fields.

Western Jumping Mouse. *Zapus princeps.* North Dakota and extreme northeastern South Dakota. Moist meadows.

Family Erethizontidae—New World Porcupines

Porcupine. *Erethizon dorsatum.* North Dakota to central Oklahoma and Texas. Near trees.

Family Myocastoridae—Nutria

Coypu or Nutria. *Myocaster coypus.* Introduced, southeastern Oklahoma. Marshes.

Family Leporidae—Hares and Rabbits

Swamp Rabbit. *Sylvilagus aquaticus.* Southeastern Kansas and eastern Oklahoma. Marshy lowlands.

Desert Cottontail. *Sylvilagus audubonii.* Western North Dakota to western Oklahoma and Texas. Arid grasslands.

Eastern Cottontail. *Sylvilagus floridanus.* Entire region. Woods, fields, farms.

Nuttall's (Mountain) Cottontail. *Sylvilagus nuttallii.* Western North Dakota and western South Dakota. Sagesteppes and pine woodlands.

Snowshoe Hare. *Lepus americanus.* North Dakota. Boreal forests.
Black-tailed Jackrabbit. *Lepus californicus.* Southern South Dakota to Oklahoma
and Texas; gradually replacing *townsendii* southward. Southern grasslands.
White-tailed Jackrabbit. *Lepus townsendii.* North Dakota to northern Kansas.
Northern grasslands.

Reptiles and Amphibians (Herptiles) of the Great Plains States

The 136 species of herptiles listed here are those occurring from North Dakota south through Oklahoma and the Texas panhandle. The vernacular English names primarily used are mostly those of Conant (1998), with some frequently encountered alternative names shown in parentheses. Habitat descriptions indicated here are minimal. Taxa within ordinal groups are arranged alphabetically by genus, and species are alphabetically listed within genera. Range statements are based on state surveys, including North Dakota (Hoberg and Gause 1992), South Dakota (Fisher et al. 1999; Ballinger, Meeker, and Thies 2000), Nebraska (Lynch 1985), Kansas (Collins 1982), Oklahoma (Webb 1970; Sievert and Sievert 1988), and Texas (Dixon 2000). Texas occurrences refer only to the panhandle region. Species endemic to or especially characteristic of the Great Plains are underlined. A few species of accidental or uncertain occurrence are excluded. "Rare," "imperiled," and "critically imperiled" are State Natural Heritage Program classifications; "threatened" and "endangered" are those of federal or state agencies.

Order Caudata—Salamanders

Three-toed Amphiuma. *Amphiuma tridactylum*. Southeastern Oklahoma. Muddy habitats.

Ringed Salamander. *Ambystoma annulatum*. Eastern Oklahoma. Pools and ponds.

Spotted Salamander. *Ambystoma maculatum*. Eastern Oklahoma. Woodland ponds.

Marbled Salamander. *Ambystoma opacum*. Southeastern Oklahoma. Varied moist habitats.

Mole Salamander. *Ambystoma talpoideum*. Southeastern Oklahoma. Damp places.

Small-mouthed Salamander. *Ambystoma texanum*. Eastern Oklahoma. Damp
  places.
Tiger Salamander. *Ambystoma tigrinum*. Entire region. Damp places.
Ouachita Dusky Salamander. *Desmognathus brimlayorum*. Southeastern Kansas.
  Wet places.
Long-tailed Salamander. *Eurycea longicauda*. Extreme southeastern Kansas
  (state threatened) and eastern Oklahoma. Damp places.
Cave Salamander. *Eurycea lucifuga*. Extreme southeastern Kansas (state endan-
  gered) and northeastern Oklahoma. Caves.
Many-ribbed Salamander. *Eurycea multiplicata*. Extreme southeastern Kansas
  (state endangered) and eastern Oklahoma. Caves.
Oklahoma Salamander. *Eurycea tynerensis*. Northeastern Oklahoma. Wet places.
Four-toed Salamander. *Hemidactylum scutatum*. Southeastern Oklahoma.
  Sphagnum bogs.
Red River Mudpuppy. *Necturus louisianensis*. Eastern Kansas and eastern Okla-
  homa. Wet places.
Mudpuppy. *Necturus maculosus*. Eastern North Dakota, northeastern South
  Dakota, and eastern Kansas. Wet places.
Eastern (Central) Newt. *Notophthalmus viridescens*. Extreme eastern Kansas
  (state threatened) and eastern Oklahoma. Swamps and ponds.
Zigzag Salamander. *Plethodon dorsalis*. Eastern Oklahoma. Damp places.
Slimy Salamander. *Plethodon glutinosus*. Eastern Oklahoma. Damp places.
Rich Mountain Salamander. *Plethodon ouachitae*. Eastern Oklahoma. Damp
  places.
Southern Redback Salamander. *Plethodon serratus*. Extreme southeastern Okla-
  homa. Damp places.
Lesser Siren. *Siren intermedia*. Southeastern Oklahoma. Shallow water.
Grotto Salamander. *Typhlotriton spelaeus*. Southeastern Kansas (state endan-
  gered) and northeastern Oklahoma. Caves.

Order Anura—Frogs and Toads
Northern Cricket Frog. *Acris crepitans*. Southeastern South Dakota to Okla-
  homa and Texas. Shallow water.
American Toad. *Bufo americanus*. Eastern North Dakota to eastern Oklahoma.
  Diverse habitats.
Great Plains Toad. *Bufo cognatus*. North Dakota to Oklahoma and Texas. Grass-
  lands.
Green Toad. *Bufo debilis*. Western Kansas (state threatened) to Texas and west-
  ern Oklahoma. Rocky areas.
Canadian Toad. *Bufo hemiophrys*. North Dakota to eastern South Dakota. Near
  water.
Red-spotted Toad. *Bufo punctatus*. Western Oklahoma and Texas. Rocky areas.

Texas Toad. *Bufo speciosus.* Western Oklahoma and Texas. Sandy soils.

Woodhouse's (Rocky Mountain) Toad. *Bufo woodhousei.* North Dakota to Oklahoma and Texas. Diverse habitats.

Great Plains Narrowmouth Toad. *Gastrophryne olivacea.* Southeastern Nebraska to Oklahoma and Texas. Moist grassy areas.

Eastern Narrowmouth Toad. *Gastrophryne carolinensis.* Extreme southeastern Kansas (state threatened) and eastern Oklahoma. Moist areas.

Green Treefrog. *Hyla cinerea.* Southeastern Oklahoma. Moist woods.

Cope's Gray Treefrog. *Hyla chrysocelis.* Southeastern South Dakota south to Oklahoma. Separable from *H. versicolor* only by voice. Moist woods.

Eastern Gray Treefrog. *Hyla versicolor.* Eastern edge of region, from extreme eastern North Dakota to southeastern Oklahoma. Moist woods.

Spotted Chorus Frog. *Pseudacris clarkii.* Central Kansas to Oklahoma and Texas. Woods near water.

Spring Peeper. *Pseudacris crucifer.* Extreme eastern Kansas (state threatened) and eastern Oklahoma. Woods near water.

Strecker's Chorus Frog. *Pseudacris streckeri.* Extreme south-central Kansas (state threatened) to eastern Oklahoma. Woods near water.

Western (Striped) Chorus Frog. *Pseudacris triseriata.* Widespread except for southwestern Kansas and western Oklahoma. Grasslands near water.

Crawfish Frog. *Rana areolata.* Eastern Kansas to eastern Oklahoma. Crawfish holes.

Plains Leopard Frog. *Rana blairi.* From southeastern South Dakota south. Plains wetlands.

Bull Frog. *Rana catesbiana.* Widespread from southern South Dakota south. Standing water.

Green Frog. *Rana clamitans.* Extreme southeastern Kansas (state threatened) and eastern Oklahoma. Brooks and small streams.

Northern Leopard Frog. *Rana pipiens.* Widespread through the Dakotas, south to the Platte River. Meadow wetlands.

Wood Frog. *Rana sylvatica.* From eastern North Dakota to northeastern South Dakota. Moist woods.

Southern Leopard Frog. *Rana utricularia.* Eastern Kansas and eastern Oklahoma. Shallow wetlands.

Plains Spadefoot Toad. *Scaphiopus (Spea) bombifrons.* North Dakota to Texas. Open grasslands.

Couch's Spadefoot. *Scaphiopus couchii.* Western Oklahoma and Texas. Arid habitats.

New Mexico (Western) Spadefoot. *Scaphiopus (Spea) multiplicatus* (previously included in *S. hammondii*). Western Oklahoma and Texas. Arid habitats.

Hurter's (Eastern) Spadefoot. *Scaphiopus holbrookii* (sometimes classified as a distinct species, *S. hurterii*). Eastern Oklahoma. Wooded areas.

Order Chelonia—Turtles

Snapping Turtle. *Chelydra serpentina.* North Dakota to Oklahoma and Texas. Diverse wetlands.

(Western) Painted Turtle. *Chrysemys picta.* North Dakota to northern Oklahoma. Diverse wetlands.

Chicken Turtle. *Deirochelys reticularia.* Southeastern Oklahoma. Still water.

Blanding's Turtle. *Emydoidea blandingi.* Central Nebraska; rare in South Dakota. Marshes.

Common Map Turtle. *Graptemys geographica.* Eastern Kansas (state threatened) and northeastern Oklahoma. Larger wetlands, especially rivers.

Mississippi Map Turtle. *Graptemys kohnii.* Eastern Kansas to eastern Oklahoma. Larger flowing wetlands.

False Map Turtle. *Graptemys pseudogeographica.* North Dakota to eastern Oklahoma. Rivers.

Yellow Mud Turtle. *Kinosternon flavescens.* Nebraska (rare) to western Oklahoma and Texas. Diverse wetlands.

Common Mud Turtle. *Kinosternon subrubrum.* Eastern Oklahoma. Shallow wetlands.

Alligator Snapping Turtle. *Macroclemys temminckii.* Eastern Kansas to eastern Oklahoma. Lakes and rivers.

Hieroglyphic River Cooter. *Pseudomys concinna.* Eastern Kansas to eastern Oklahoma. Rivers.

Common Musk Turtle. *Sternotherus odoratus.* Eastern Kansas to eastern Oklahoma. Wooded swamps.

Razorback Musk Turtle. *Sternotheru carinus.* Eastern Oklahoma. Streams and swamps.

Three-toed (Common) Box Turtle. *Terrapena carolina trianguis.* Eastern Kansas to eastern Oklahoma. Woodlands.

Ornate (Western) Box Turtle. *Terrapena ornata.* South Dakota to Oklahoma and Texas. Open plains, especially grasslands.

Common (Red-eared) Slider. *Trachemys (Pseudemys) scripta.* Extreme southeastern Nebraska, and from Kansas to Oklahoma and Texas. Diverse wetlands.

Smooth Softshell. *Apalone (Trionyx) mutica.* North Dakota to Oklahoma and Texas. Rivers.

Spiny Softshell. *Apalone (Trionyx) spinifera.* Southeastern South Dakota and Nebraska to Oklahoma and Texas. Rivers.

Order Lacertilia—Lizards

Green Anole. *Anolis carolinensis.* Southeastern Oklahoma. Diverse terrestrial habitats.

Texas Spotted Whiptail. *Cnemidophorus gularis.* Southern Oklahoma and Texas. Dry grasslands.

Six-lined Racerunner. *Cnemidophorus sexlineatus.* South Dakota (rare) to Oklahoma and Texas. Open areas with sandy or loose soils.

Checkered Whiptail. *Cnemidophorus tesselatus.* Western Oklahoma and Texas. Rocky grasslands.

Collared Lizard. *Crotophytus collaris.* Kansas, Oklahoma, and Texas. Rocky areas.

Coal Skink. *Eumeces anthracinus.* Eastern Kansas to eastern Oklahoma. Wooded hillsides.

Five-lined Skink. *Eumeces fasciatus.* Extreme southeastern Nebraska, eastern Kansas, and eastern Oklahoma. Damp areas near woods.

Broadhead Skink. *Eumeces laticeps.* Eastern Kansas (state threatened) and eastern Oklahoma. Wooded habitats.

Many-lined Skink. *Eumeces multivirgatus.* South Dakota (rare) and Nebraska. Open plains and sandhills.

Great Plains Skink. *Eumeces obsoletus.* South Dakota to Oklahoma and Texas. Grasslands.

Prairie Skink. *Eumeces septentrionalis.* Eastern North Dakota to Oklahoma. Sandy or gravelly habitats.

Speckled (Lesser) Earless Lizard. *Holbrookia maculata.* Southern South Dakota (rare) to Oklahoma and Texas. Arid or sandy plains.

Slender Glass Lizard. *Ophisaurus attenuatus.* Southeastern Nebraska to Oklahoma. Dry grasslands and open woods.

Texas Horned Lizard. *Phrynosoma cornutum.* Kansas to Oklahoma and Texas; declining. Open areas, rocky or sandy soils.

Short-horned Horned Lizard. *Phrynosoma douglassi* (now often called *P. hernandezii*). North Dakota to western Nebraska. Short-grass plains.

Round-tailed Horned Lizard. *Phrynosoma modestum.* Western Oklahoma and Texas. Arid, scrub habitats.

Sagebrush Lizard. *Sceloperus graciosus.* Western North Dakota and western Nebraska. Sagesteppes.

Northern Prairie Lizard. *Sceloperus undulatus.* Southern South Dakota to Oklahoma and Texas. Prairies, sandy habitats.

Ground Skink. *Scincella (Lygosoma) lateralis.* Eastern Kansas and Oklahoma. Wooded habitats.

Side-blotched Lizard. *Uta stansburiana.* Southwestern Oklahoma and Texas. Arid and sandy areas.

Order Crocodylia—Alligators and Crocodiles
American Alligator. *Alligator mississippiensis.* Southeastern Oklahoma. Swamps.

Order Serpentes—Snakes

Copperhead Snake. *Agkistrodon contortrix*. Southeastern Nebraska, eastern Kansas, and Oklahoma. Rocky wooded hillsides.

Cottonmouth. *Agkistrodon piscivorus*. Eastern Oklahoma. Near water.

Glossy (Faded) Snake. *Arizona elegans*. Southwestern Nebraska to Oklahoma and Texas. Sandy areas.

Eastern Worm Snake. *Carphophis amoenus*. Southeastern Nebraska to eastern Oklahoma. Moist habitats.

Scarlet Snake. *Cemophora coccinea*. Eastern Oklahoma. Sandy or other soft-substrate habitats.

Blue Racer. *Coluber constrictor*. Western North Dakota to Oklahoma and Texas. Open habitats.

Western Diamondback Rattlesnake. *Crotalus atrox*. Oklahoma and Texas; very rare in extreme southern Kansas. Arid habitats.

Timber Rattlesnake. *Crotalus horridus*. Southeastern Nebraska to eastern Oklahoma. Wooded areas.

Prairie (Western) Rattlesnake. *Crotalus v. viridis*. North Dakota to Oklahoma and Texas. Prairies and plains.

Ringneck Snake. *Diadophis punctatus*. Southeastern South Dakota to Oklahoma and Texas. Wooded areas.

Corn Snake. *Elaphe guttata*. Southern Nebraska to Oklahoma and Texas. Diverse habitats.

Black (Pilot) Rat Snake. *Elaphe obsoleta*. Southeastern Nebraska to Oklahoma. Diverse habitats.

Fox Snake. *Elaphe vulpina*. Southeastern South Dakota and eastern Nebraska. Diverse habitats.

Mud Snake. *Farancia abacura*. Southeastern Oklahoma. Swamps and lowlands.

Western Hognose Snake. *Heterodon nasicus*. North Dakota to Oklahoma and Texas. Open, often sandy areas.

Eastern Hognose Snake. *Heterodon platyrhinos*. Southeastern South Dakota (rare) to Oklahoma and Texas. Sandy areas.

Night Snake. *Hypsiglena torquata*. Southern Kansas (state threatened), Oklahoma, and Texas. Arid, rocky habitats.

Prairie Kingsnake. *Lampropeltis calligaster*. Southeast Nebraska to Oklahoma and Texas. Prairies and savannas.

Eastern (Common) Kingsnake. *Lampropeltis getula*. Southeastern Nebraska to Oklahoma and Texas. Near water.

(Red) Milk Snake. *Lampropeltis triangulum*. South Dakota (rare) to Oklahoma and Texas. Diverse habitats.

Texas Slender Blind Snake. *Leptotyphlops dulcis*. Southern Kansas (state threatened) to Oklahoma and Texas. Arid plains.

Western Coachwhip Snake. *Masticophis flagellum.* Southwestern Nebraska to Oklahoma and Texas. Open grasslands and brushlands.

Plain-bellied Water Snake. *Nerodia erythrogaster.* Kansas to Oklahoma and Texas. Near water.

Southern Water Snake. *Nerodia fasciata.* Southeastern Oklahoma. Near water.

Diamondback Water Snake. *Nerodia rhombifera.* Kansas to Oklahoma and Texas. Near water.

Common (Northern) Water Snake. *Nerodia (Natrix) sipedon.* Nebraska to Oklahoma. Near water.

Rough Green Snake. *Opheodrys aestivis.* Eastern Kansas and eastern Oklahoma. Dense vegetation near water.

Smooth Green Snake. *Opheodrys vernalis.* North Dakota to Nebraska. Critically imperiled in Nebraska, imperiled in Black Hills (endemic population). Grasslands.

Bullsnake. *Pituophis catenifer* (including *melanoleucus*). Western North Dakota to Oklahoma and Texas. The form *melanoleucus* sometimes is specifically recognized as the "Pine Snake." Prairies and plains.

Graham's Crayfish (Water) Snake. *Regina grahamii.* Eastern Nebraska to Oklahoma and Texas. Near water.

Glossy Crayfish Snake. *Regina rigida.* Southeastern Oklahoma. Near water.

Long-nosed Snake. *Rhinocheilus lecontei.* Southwestern Kansas (state threatened; critically imperiled), Oklahoma, and Texas. Arid habitats.

Massasauga Rattlesnake. *Sistrurus catenatus.* Southeastern Nebraska to Oklahoma and Texas; very rare in Nebraska and Kansas (imperiled). Moist prairies.

Pygmy Rattlesnake. *Sistrurus miliarus.* Eastern Oklahoma. Near water.

Great Plains Ground Snake. *Sonora semiannulata* (= *episcopa*). Kansas to Oklahoma and Texas. Open plains.

Brown Snake. *Storeria dekayi.* Southern Nebraska to Oklahoma and Texas; very rare in South Dakota. Woods, other varied habitats.

Red-bellied Snake. *Stoereia occipitomaculata* Eastern North Dakota to Nebraska (critically imperiled), eastern Kansas (state threatened), and Oklahoma. State threatened in South Dakota; Black Hills race *pahasapae* critically imperiled. Near open woods.

Flat-headed Snake. *Tantilla gracilis.* Eastern Kansas to Oklahoma and Texas. Rocky areas.

Plains Black-headed Snake. *Tantilla nigriceps.* Southwestern Nebraska to Oklahoma and Texas. Rocky areas.

Western Terrestrial (Wandering) Garter Snake. *Thamnophis elegans.* Western South Dakota, northwestern Nebraska, and western Oklahoma. Diverse habitats, often near water.

Checkered Garter Snake. *Thamnophis marcianus.* Southern Kansas (state threatened) to Oklahoma and Texas. Near water.

Western Ribbon Snake. *Thamnophis proximus.* Eastern Nebraska to Oklahoma and Texas. Near water.

Plains Garter Snake. *Thamnophis radix.* North Dakota to western Oklahoma and Texas. Diverse habitats, often near water.

Common Garter Snake. *Thamnophis sirtalis.* North Dakota to Oklahoma and Texas. Diverse habitats, often near water.

Lined Snake. *Tropicoclonion lineatum.* Eastern South Dakota (no recent records) to Oklahoma and Texas. Open prairies, sparsely timbered habitats.

Rough Earth Snake. *Virginia striatula.* Southeastern Kansas and eastern Oklahoma. Moist soils in or near forests.

Smooth Earth Snake. *Virginia valeriae.* Eastern Kansas (state threatened) and eastern Oklahoma. Moist soils in or near forests.

Other Animals and Plants Mentioned in the Text

Arctic fox. *Alopex lagopus.*
Arctic hare. *Lepus arcticus.*
Aromatic sumac. *Rhus aromatica.*
Astrapias (birds-of-paradise). *Astrapia* spp.
Bald cypress. *Taxodium distichum.*
Big bluestem. *Andropogon gerardii.*
Big sagebrush. *Artemisia tridentata.*
Blowoutgrass. *Redfieldia flexuosa.*
Blowout penstemon. *Penstemon haydenii.*
Blue grama. *Bouteloua gracilis.*
Blue grouse. *Dendragapus obscurus.*
Brant (goose). *Branta bernicla.*
Buffalo grass. *Buchloe dactyloides.*
Bur oak. *Quercus macrocarpa.*
Columbian mammoth. *Mammuthus columbi.*
Common cuckoo. *Cuculus canorus.*
Common (Eurasian) crane. *Grus grus.*
Creeping juniper. *Juniperus horizontalis.*
Daisy fleabane. *Erigeron strigosus.*
Dwarf sage. *Artemisia nana.*

Eastern cottonwood. *Populus deltoides.*
Eastern red cedar. *Juniperus virginiana.*
Eiders (ducks). *Somateria* spp.
Elms. *Ulmus* spp.
Eurasian crane. *Grus grus.*
Evening primroses. *Oenothera* spp.
Giant short-faced bear. *Arctidus femus.*
Grama grasses. *Bouteloua* spp.
Gray flycatcher. *Empidonax wrightii.*
Great Plains yucca. *Yucca glauca.*
Green-tailed towhee. *Pipilo chlorurus.*
Grizzly bear. *Ursus arctos.*
Gunnison sage-grouse. *Urophasianus minimus.*
Hackberry. *Celtis occidentalis.*
Hairy grama. *Bouteloua hirsuta.*
Hayden's (Blowout) penstemon. *Penstemon haydenii.*
Hickories. *Carya* spp.
House mouse. *Mus musculus.*
Indian grass. *Sorghastrum nutans.*

Jaegers. *Stercorarius* spp.
Leadplant. *Amorpha canescens.*
Lemmings. *Lemmus* and *Dicrostonyx* spp.
Limber pine. *Pinus flexilis.*
Little bluestem. *Schizachyrium scoparium.*
Lodgepole pine. *Pinus contorta.*
Long-leaved sage. *Artemisia longifolia.*
Mallows. *Sphaeralcea* spp.
Mesquite. *Prosopsis juliflora.*
Mexican ground squirrel. *Spermophilus mexicanus.*
Milk vetches. *Astragalus* spp.
Mountain mahogany. *Cercocarpus montanus.*
Needle-and-thread. *Stipa comata.*
Norway rat. *Rattus norvegicus.*
Oaks. *Quercus* spp.
Paper birch. *Betula papyrifera.*
Pasque flower. *Anemone patens.*
Pines. *Pinus* spp.
Pinyon pine. *Pinus edulis.*
Polar bear. *Thalarctos maritimus.*
Ponderosa pine. *Pinus ponderosa.*
Prairie rose. *Rosa arkansana.*
Prickly pear cactus. *Opuntia* spp.
Ptarmigans. *Lagopus* spp.

Purple avens. *Geum triflorum.*
Pygmy rabbit. *Sylvilagus idahoensis.*
Quaking aspen. *Populus tremuloides.*
Rocky Mountain (Western) red cedar. *Juniperus scopulorum.*
Russian olive. *Elaeagnus angustifolia.*
Sagebrush lizard. *Sceloperus graciosus.*
Sand cherry. *Prunus besseyi.*
Sand dropseed. *Sporobolus cryptandrus.*
Sand lovegrass. *Eragrostis trichodes.*
Sand muhly. *Muhlenbergia pungens.*
Sand sagebrush. *Artemisia filifolia.*
Sandhills bluestem. *Andropogon hallii.*
Shinnery oak. *Quercus havardii.*
Side-oats grama. *Bouteloua curtipendula.*
Silver sagebrush. *Artemisia cana.*
Skunkbrush (Aromatic) sumac. *Rhus aromatica.*
Tupelo. *Nyssa aquatica.*
Western red cedar. *Juniperus scopulorum*
White spruce. *Picea glauca.*
Willow ptarmigan. *Lagopus lagopus.*
Woolly mammoth. *Mammuthus primigenius.*

GLOSSARY

Adaptation. An evolved structural, behavioral, or physiological trait that increases an organism's individual fitness (its ability to survive and reproduce).

Advertising behavior. The social behaviors (or "displays") of an animal of either sex that may serve to identify and announce its species, sex, reproductive state, individual social status, and overall vigor. *See also* mechanical signals, pheromones, vocalizations.

Aestivation. *See* estivation.

Agonistic. The entire dominance/submissive behavioral spectrum, comprising a response gradient ranging from attack to escape, including intermediate ambivalent states. *See also* signals.

Alkaline. Liquids or solids that are bitter and sometimes caustic owing to the variable presence of soluble mineral salts.

Alluvial. Moved or deposited by water. *See also* eolian.

Altricial. Referring to the condition of being hatched or born in a relatively undeveloped stage, usually both blind and helpless. *See also* nidicolous.

Anterior. Toward or at the front.

Aquifer. A subterranean zone of saturated sands, gravels, and so on, composing the "water table."

Arboreal. Tree-dwelling.

Arborescent. Treelike.

Arenicolous. Adapted to life in sand.

Arthropod. A member of the invertebrate phylum Arthropoda, which includes insects, crustaceans, and other "jointed-legged" animals with chitinous or calcareous exoskeletons. *See also* invertebrate.

Association. A specific type of biotic community, usually named for one or more plant species or genera that consistently occur as climax dominants within that community. *See also* climax community, dominant.

Avian. Relating to birds.

Avifauna. The collective bird life of a particular locality or region.

275

Badlands. Vernacular term for highly eroded landforms, best developed in arid and unglaciated parts in the northern Great Plains, such as along the White, Cheyenne, and Little Missouri Rivers.

Biome. A major regional ecosystem, including both the plants and animals. Comparable to an "ecoregion" if defined mainly by landscape geography rather than floristically. *See also* ecoregion, formation.

Blowout. A local area of bare, eroding sand in an otherwise vegetated dune.

Boreal. Having northern affinities, such as the coniferous forests of Canada.

Boreomontane. Having a distribution that includes both northerly and mountainous regions.

Braided stream. A meandering stream having several shallow and interconnected channels.

Brood. As a verb, to cover and apply heat to hatched young (incubation is the same behavior when applied to eggs). Also refers (as a noun) to a group of siblings typically tended by a single female or pair. *See also* litter, siblings.

Brood parasitism. The behavior of a species in which females lay their eggs in the nests of their own species (intraspecific brood parasitism) or those of other species (interspecific brood parasitism). Also called nest parasitism and egg parasitism.

Browse. To forage on the twigs and leaves of woody plants. "Browse" is also sometimes used to refer to the vegetation itself. *See also* graze.

Bunchgrasses. Perennial grasses that grow and expand laterally through stolons or rhizomes, forming distinct, discontinuous clumps rather than the relatively continuous ground cover typical of sod-forming grasses.

Caliche. Deposits of calcium carbonate (hardpan) that accumulate in the soils of arid regions. *See also* caprock escarpment.

Calls. Vocalizations that are acoustically simple and are not usually limited by age, sex, or season. *See also* signals, songs, vocalizations.

Campestrian. Living on the plains.

Candidate. A term used by the Office of Endangered Species in identifying those taxa for which substantial information exists to support a proposal to list them as threatened or endangered (Category 1), or for which the population status is under investigation (Category 2). Category 3 species are those that are no longer under consideration for such listing for various reasons.

Canid. A member of the dog family Canidae.

Caprock escarpment. An eroded escarpment of limestone (caliche) in the central panhandle of Texas. *See also* caliche, escarpment.

Carnivore. An animal that consumes meat, especially that which it has killed itself, as opposed to scavengers that eat carrion. *See also* omnivore.

Carotenoids. Red to yellow pigments derived from plant or animal materials, and often present in feathers or scales. *See also* melanins.

Caudal. At or toward the tail.

Cavernicolous. Refers to living in caves.

Centimeter. A metric measurement equal to 0.39 inch; also represents ten millimeters.

Clay. Fine-grained earth materials that have particle sizes forming colloidal suspensions or pastes in water and highly adherent solids when dry and heated. *See also* silt.

Climax community. A stable aggregation of locally interacting species that is self-perpetuating and (at least measurably) is no longer undergoing ecological succession. *See also* community, dominant, ecoregion, formation, succession.

Cline. A gradient in some structure or trait in geographically separated populations.

Clutch. A set of eggs (one or more), normally those laid by a single female and incubated simultaneously.

Commensal. A species that lives in close association with another and gains some benefits from it without doing the former any direct harm.

Community. In ecology, an interacting group of plants, animals, and microorganisms situated in a specific location. *See also* ecosystem, habitat.

Congeneric. Belonging to the same genus.

Coniferous. Cone-bearing; conifers are cone-bearing, usually nondeciduous trees.

Conspecific. Belonging to the same species.

Cool-season grasses. Grasses adapted for growing in cooler climates, and requiring more water than warm-season grasses. Also called C_3 grasses in reference to an intermediate three-carbon molecule that is present during photosynthesis. *See also* warm-season grasses.

Cordilleran. Pertaining to mountains.

Cosmopolitan. Widely or universally distributed.

Coteau. A sloping or hilly area, such as the Missouri Coteau of the Dakotas.

Courtship. *See* advertising behavior, mating, signals.

Crepuscular. Associated with dawn and dusk, occurring between diurnal (daytime) and nocturnal (nighttime) activity patterns.

Cursorial. Running.

Deciduous. Descriptive of shedding, especially of trees that drop their leaves after each growing season, but also used in reference to nonpermanent teeth.

Dichromatism. The presence of two distinct and genetically controlled plumage/pelage patterns or colors among adults of a species' population (sometimes called "phases"). *See also* dimorphism, morphism.

Dimorphism. The presence of two distinct forms (morphs) that differ absolutely or statistically within a single population in one or more measurable traits. *See also* dichromatism, morphism.

Display. An evolved ("ritualized") behavior that communicates information within or between species. *See also* advertising behavior, ritualization, signals, territoriality.

Diurnal. Active during the daytime. *See also* crepuscular, nocturnal.

Dominant. In ecology, descriptive of plant taxa that exert the strongest ecological effects (control of energy flow) within a community. *See also* community, keystone species.

Dorsal. Referring to the back or upper surface. *See also* ventral.

Drift. Glacially transported and deposited materials, either unsorted by size (till) or size-sorted (stratified drift). *See also* glacial erratic, moraine, till.

Ear conch. The enlarged outer ear area of owls and many mammals, variously specialized for receiving faint sounds.

Echolocation. The use of high-frequency sounds by some mammals to detect prey or to navigate in the dark.

Ecoregions. Areas having similar climates, geomorphology, and potential natural vegetation composed of clusters of interacting landscapes. *See also* biome, formation.

Ecosystem. An interacting group of plants, animals, microorganisms, and their physical environment of no specific size.

Ecotone. An ecologic transition zone that physically connects two quite different biotic communities. Ecotones may be broad or narrow and relatively stable or dynamically variable through time. *See also* edge species, sere.

Edge-effect. The tendency for transitional habitats, such as forest edges or shorelines, to exhibit high species diversity, often supporting species typical of both the adjoining habitat types and sometimes also unique ones.

Edge species. A species that is more common in or most characteristic of transitional communities, such as forest-edge species. *See also* ecotone.

Endangered. Descriptive of taxa existing in such small numbers as to be in direct danger of extinction without human intervention. Some species have been formally identified as "endangered" or "threatened" by federal or state agencies and are given special protection. *See also* candidate, threatened.

Endemic. Descriptive of taxa that are both native to and limited to a specific area, habitat, or region, such as the Great Plains. *See also* indigenous.

Environment. The natural surroundings of an organism or community of organisms. *See also* habitat.

Eolian (or aeolian). Shaped, carried, or influenced by wind; for example, wind-produced sounds such as whistles and wind-carried soil particles such as silt. *See also* alluvial, loess.

Ephemeral. Short-lived phenomena. *See also* playa lake, vernal pond.

Epigamic. Sexual, as in epigamic signals.

Escarpment. A steep, eroded slope; also popularly termed a "scarp."

Estivation (also spelled aestivation). A period of dormancy during summer. *See also* hibernation.

Estrus. The period of time during which a female mammal is receptive to mating. *See also* rut.

Evolution. Any gradual change. Biological or organic evolution results from changing gene frequencies in successive generations that are associated with biological adaptation, typically as a result of natural selection. *See also* natural selection.

Extinct. An organism that no longer exists anywhere. *See also* extirpated.

Extirpated. An organism that has been eliminated from some part of its range but still exists elsewhere. *See also* extinct.

Fen. A wetland characterized by having a substrate of organic matter (often peat) but, unlike typical bogs, having favorable plant nutrition levels and much greater organic productivity. *See also* marsh, peat.

Feral. Pertaining to a domestic animal that has reverted to the wild.

Fidelity. A tendency of an individual to return for breeding to a prior nest site (nest-site fidelity or tenacity), a prior year's mate (mate fidelity), or the individual's area of rearing (natal fidelity or philopatry). *See also* tenacity.

Fledging. The initial acquisition of flight by a bird. The period from hatching to fledging is called the fledging period, which often corresponds to the nestling period or is slightly longer. *See also* nestling.

Fledgling. A newly fledged juvenile bird. *See also* juvenile.

Floodplain. The low, relatively flat area of a river valley that is subjected to occasional flooding.

Food chain. The sequence of energy transformations in nature, from primary "producers" (green plants) through a series of plant-eaters (herbivores) and meat-eaters (carnivores), collectively called "consumers." "Decomposers" such as bacteria recycle organic matter back to its inorganic components.

Forest. A general term for a community dominated by rather tall trees, in which the height of the trees (typically more than about 20 feet) is much greater than the average distance between them, and the overhead canopy is more or less continuous. *See also* savanna, woodland.

Formation. In ecology, a major type of plant community (or "biome," if the animals are included) that extends over broad regions that collectively share similar climates, soils, biological succession patterns, and have similar life forms of dominant plants at their eventually stable or "climax" vegetational stage. *See also* association.

Fossorial. Digging.

Fratricide. The killing of a nestmate, usually the younger or weaker members of a brood; also called siblicide.

Fuscous. Brownish black to grayish brown.

Gallery forest. Narrow riverine forests that follow waterways out into otherwise nonforested habitats. *See also* riparian, riverine.

Genus (adj., generic; pl., genera). A "general" Latin or Latinized name that is applied to plants and animals. The genus is the first (and always capital-ized and italicized) component of a species' usually two-part scientific name. *See also* species.

Gestation period. The time required between fertilization and birth in pla-cental mammals. *See also* incubation period.

Glacial erratics. Large rocks and boulders that have been glacially transported and randomly deposited over the landscape, having a composition differ-ent from that of the bedrock below. *See also* drift, moraine.

Gram. A metric unit of weight; 28.3 grams equal an ounce, and 454 grams equal a pound.

Granivore. Grain- or seed-eating; granivores typically eat seeds of grasses and other nonwoody plants. *See also* herbivore, mast.

Grass. Herbaceous plants distinguished by having a (usually) hollow stem, narrow leaves that are parallel-veined, and tiny flowers and seeds borne on small spikes. *See also* sedge.

Graze. To feed on the leaves of herbaceous plants. *See also* browse.

Great Plains. The nonmountainous region of interior North America lying east of the Rocky Mountains and west of the Central Lowlands of the Mis-sissippi and lower Missouri drainages. The "central plains" is a convenient inclusive term encompassing the Central Lowlands. *See also* high plains.

Groveland. A mostly nonforested community, but with scattered forestlike clusters of trees, especially aspen groves, common in the grassland–boreal forest ecotone area of northern North Dakota. *See also* parkland, forest.

Guild. A group of species that exploits the same class of environmental char-acteristics (such as food types) in a similar way, whether or not they are closely related. *See also* niche.

Habitat. The ecological situation or community type in which a species sur-vives; its natural "address." *See also* community, ecosystem, niche.

Halophytes. Plants tolerant of high levels of environmental salts; often also called xerophytic (drought-tolerant) plants.

Hectare (abbreviated ha). A metrically defined area (10,000 square meters) equal to 2.47 acres. A square mile contains 259 hectares, or 640 acres.

Herb. Any plant with no permanent above-ground parts, including grasses, sedges, rushes, and forbs. *See also* forb.

Herbivore. An animal that mainly consumes soft plant materials (grasses and other herbs). *See also* granivorous.

Herptile (or herp). A member of the amphibian and reptile vertebrate assem-blage. Herpetology refers to the study of such animals, and the herpeto-fauna includes the reptiles and amphibians of a region.

Hertz (Hz). A unit of sound frequency equal to one cycle per second (cps); kilohertz (kHz) units equal 1,000 cps.

Hibernation. A period of winter dormancy. *See also* estivation.

Hispid. Covered with stiff hairs.

Home range. The entire area used by an individual, pair, or family over a specified period, such as a year or lifetime. *See also* territory.

Host. Among birds, those species that receive and accept eggs of brood parasites; also used for more generalized host-parasite relationships.

Imperiled. A term used by the Natural Heritage Program of the Nature Conservancy to identify species in clear danger of extinction. The highest classification level is "critically imperiled." *See also* endangered.

Incubation. In birds, the parental application of body heat to eggs; also the development of an embryo within an enclosed egg. The incubation period is the time required between the start of incubation and hatching. *See also* brooding, gestation period.

Indigenous. Descriptive of taxa that are native to, but not necessarily limited to, a particular area or region. *See also* endemic.

Innate. Pertaining to genetically transmitted traits, especially behavioral ones.

Insectivorous. Having a diet composed mostly of insects and other arthropods.

Instinct. An innate behavioral trait that appears to be more complex than various simpler innate responses such as reflexes.

Integument. The surface covering of an animal, such as skin, fur, or feathers.

Intergeneric. Pertaining to interactions between genera.

Interspecific. Pertaining to interactions between species.

Intraspecific. Pertaining to interactions within species.

Invertebrates. Animals lacking backbones, but often with exoskeletons, such as arthropods. *See also* vertebrates, arthropods.

Irruption. A periodic large-scale movement of an animal population into an area. *See also* migration.

Juvenal. The feathers (or collective plumage) acquired by birds during the nestling period, which are carried for a variable time after fledging (initial flight).

Juvenile. A sexually immature or subadult stage in an animal's life. Among birds, the feathers carried are predominantly those of the juvenal plumage. *See also* juvenal.

Kettle. A name applied to a group of flying hawks aggregated in a migrating group.

Keystone species. An animal species whose presence in a community has a significant effect on the community's overall ecological structure. *See also* dominant.

Kilogram (kg). A metric measure equal to 1,000 grams, or about 2.2 pounds.

Kilometer (km). A metric distance of 1,000 meters, or 0.62 mile. A square kilometer equals 0.386 square mile.

Lagoon. A term popularly used in the central plains to refer to temporary spring wetlands that depend on seasonal precipitation for annual recharge. *See also* playa lake, pothole, vernal pond.

Lake. A freshwater or saline wetland that is typically large enough (usually over 40 acres) to have barren, wave-washed shorelines and deep enough (usually more than 6 feet) to develop seasonal thermal stratification. *See also* marsh, playa lake.

Larva (plural, larvae). Refers to immature growth stages of fish, amphibians, and many invertebrates that have not yet undergone metamorphosis into their adult form.

Larynx. The vocal organ of mammals and amphibians, located in the upper trachea. *See also* syrinx.

Lateral. Away from the midline, toward the sides.

Latin name. The (usually) two-part (generic plus specific) Latin or Latinized name given a species when it is first formally described, and by which it thereafter is properly identified; thus the species' "scientific name." *See also* genus, species, vernacular name.

Legumes. A family of plants (Leguminaceae or Fabiaceae), many of which harbor bacteria in their roots that can convert gaseous nitrogen into a molecular form usable by other plants.

Lek. A Scandinavian word used for describing a site at which communal display occurs among males of various nonmonogamous birds in competition for mating rights. As a verb, "lekking" is sometimes also used to refer to the collective social behavior of the participating males.

Life form. A term broadly descriptive of plant categories, such as coniferous or deciduous trees (or forests), broad-leaved shrubs, perennial grasses, annual forbs, and so on.

Litter. Dead, nonwoody plant matter lying on the ground surface. The term is also used to refer to the offspring produced as the result of a single gestation cycle in mammals. *See also* thatch.

Loam (or loamy). Refers to soils containing a mixture of particle sizes, including sand, silt, and clay. *See also* clay, loess, sand, till.

Loess (or loessal) **soils.** Silty soils that have been transported and deposited by wind. These soils show little or no vertical stratification and are easily eroded. The word is German (meaning "loose"), and is pronounced "luss." *See also* eolian, loam, till.

Marsh. A wetland type in which the soil is saturated for long periods of time and in which peat does not accumulate. Large marshes often grade into or form the edges of lakes. Great Plains marshes, excepting the most alkaline ones, typically have extensive shoreline vegetation and are sufficiently shallow that emergent aquatic plants cover most or all of their surface. *See also* fen, lake.

Mast. A collective term for nuts.

Mating. An indefinite, nontechnical term that, like "courtship behavior," is often loosely applied to either (1) initial pair-bonding or (2) copulation. Mat-

ing systems are types of evolved sexual relationships associated with reproduction. *See also* advertising behavior, pairing, signals.

Mechanical signals. Nonvocal social signals, including striking (percussion)
sounds such as foot-stamping, scraping (stridulation) sounds such as feathers rubbing together or on the ground, and vibration (eolian) sounds such
as those made by feathers during flight. *See also* advertising behavior, signals, vocalizations.

Melanins. Skin, hair, scale, and feather pigments of mostly gray to brown or
black hues. Melanins are common to many vertebrates and help absorb
harmful UV rays. *See also* carotenoids.

Mesic. A habitat with a moderate level of soil moisture (or other general environmental conditions, such as temperature extremes) between extremes of
xeric (dry) and hydric (wet). *See also* xeric.

Metamorphosis. Refers to organisms that undergo major bodily transformations during their lifetimes, such as most insects and amphibians.

Meter. A metric measurement equal to 100 centimeters, 1.094 yards, or 39.37
inches.

Microtine (synonym, arvicoline). Refers to a group (subfamily) of rodents that
includes voles, lemmings, and muskrats. The word is derived from the vole
genus *Microtus*. *See also* vole.

Midden. A refuse heap.

Migration. The regular, usually seasonal movements of animals between two
locations, usually breeding and nonbreeding areas. Irregular or one-way
movements may include dispersals, emigrations, immigrations, and irrup-
tions. *See also* irruption.

Mixed-grass (or midgrass) **prairie.** Perennial grasslands that are dominated
by grasses of intermediate heights (often from 1.5 to 3.0 feet tall at maturity).
Mixed-grass prairies usually are situated geographically between tallgrass
and short-grass prairies. *See also* short-grass prairie and tallgrass prairie.

Molt. The periodic loss and replacement of bodily parts or integument during a lifetime or a season. Molting in birds involves feather replacement,
producing a new plumage; in mammals, hair replacement produces a new
pelage. Among arthropods, molting refers to the periodic shedding of the
exoskeleton during growth and maturation.

Monogamy. A pair-bonding system characterized by a single male and female
remaining together for part or all of a breeding season or sometimes indefinitely. *See also* polygyny, promiscuity.

Montane. Mountainous.

Moraine. Gently rolling landscapes of glacial drift deposits. *See also* drift, till.

Morphism. An example of any of the variant types of appearance ("morphs"
or "phases") that occur in a species, including both sex-related or nonsex-
ual morphism. *See also* dichromatism, dimorphism.

Mustelid. A member of the mammalian family Mustelidae, including weasels, ferrets, minks, otters, and badgers.

Natal plumage. The initial feather covering of a newly hatched bird, often downy in most birds.

Natural selection. Differential survival and reproduction of the fittest individuals within natural populations. *See also* sexual selection.

Neotropical migrants. Those migratory birds that winter in the Neotropic Region, namely from Mexico and Central America southward. Some of these are additionally transequatorial migrants, wintering south of the Equator.

Nestling. A recently hatched bird that is still confined to the nest. *See also* fledgling.

Niche. The behavioral, morphological, and physiological adaptations of a species to its habitat. *See also* community, habitat.

Nidicolous. Those bird species whose young are hatched in a helpless condition, are reared in the nest, and need prolonged parental brooding and feeding. *See also* altricial, nestling.

Nocturnal. Active at night. *See also* crepuscular, diurnal.

Nomenclature. The process of naming objects; binomial nomenclature provides a standardized two-part Latin or Latinized name for organisms, the generic (general) name or genus followed by a specific epithet. *See also* genus, Latin name, species.

Nuptial plumage. The definitive breeding plumage of adult birds. Now generally termed "alternative plumage" in North America.

Omnivore. An animal that consumes a wide variety of plant and animal food. *See also* carnivore, granivore, herbivore.

Ornithology. The scientific study of birds.

Ovoviviparous. Refers to the production (in some reptiles) of shelled eggs that hatch within the female's reproductive tract, so that the young are born alive.

Pair-bond. The establishment of a variably prolonged social bond between two individuals, typically for facilitating reproduction. *See also* mating.

Pairing. The behavior patterns related to (1) mate choice and associated pair formation, and (2) pair-bond maintenance. Sometimes also called mating, which term is often applied to copulation. *See also* advertising behavior, mating.

Parkland. A mostly forested community, with scattered, nonforested meadowlike areas, or "parks," present in the forest matrix. *See also* groveland.

Partners in Flight. An international governmental and nongovernmental organization devoted to bird conservation in the Western Hemisphere.

Passerine. Descriptive of members of the avian order Passeriformes; popularly also called "perching birds" (because of their long hind toes) or "songbirds" (because of their typically complex vocal structures).

Peat. Undecomposed organic matter that accumulates at the bottoms of some wetlands, especially bogs and fens. *See also* fen.

Pelage. The collective hair and fur coat of a mammal. *See also* molt, plumage.

Phenotype (adj., phenotypic). The general appearance of an individual, reflecting both its genetic makeup (genotype) and possible age or environmental influences.

Pheromones. Odorous materials produced and released into the air or water by one individual and perceived by another individual of the same species, evoking a behavioral or physiological effect.

Philopatry. The tendency of an individual to return to a particular location. Natal philopatry refers to a return to one's place of hatching or birth and rearing. *See also* site tenacity.

Photosynthesis. The metabolic process by which gaseous carbon dioxide and plant-supplied water are converted into simple sugars using light energy in the presence of chlorophyll.

Piedmont. A zone or region of mountain foothills.

Plains. A nontechnical descriptive term for a flatland, especially a nonforested flatland. The "high plains" of western North America are the arid uplands east of the Rocky Mountains that support native short-grass and shrubsteppe vegetation. *See also* prairie.

Playa lake. An ephemeral, shallow lake in a wind-eroded basin, typically dependent upon irregular or seasonal rains for its existence. *See also* pluvial lake.

Pleistocene epoch. The interval extending from the end of the Pliocene epoch, or about 1.77 million years ago (mya), to 11,000 years ago. The Pleistocene includes four major glacial periods, named after their southernmost geographic limits, of which the Nebraskan glaciation (about 2.0 to 1.75 mya) was the earliest and the Wisconsinian glaciation (about 150,000 to 15,000 years ago) the most recent. The Kansan (about 1.4 to 0.9 mya) and Illinoian (about 0.55 to 0.35 mya) glaciations were of intermediate age.

Plumage. The collective feather coat of a bird. *See also* molt, pelage.

Pluvial lake. A now-extinct lake that was formed during an earlier time from higher rainfall or glacial melting. *See also* playa lake.

Polyandry. A nonmonogamous avian mating system, in which a female may temporarily associate and copulate with two or more males, typically laying a clutch of eggs for each of them to incubate and rear. *See also* polygyny, promiscuity.

Polygyny. A nonmonogamous vertebrate mating system, in which a male may temporarily associate and copulate with two or more females, either simultaneously (harem polygyny) or successively (serial polygyny). *See also* monogamy, promiscuity.

Posterior. Toward or at the rear.

Pothole. A popular term used in the northern plains to refer to small, marshy wetlands in areas of undulating glacial till that often depend on winter and spring precipitation for annual recharge. *See also* marsh, playa lake, vernal pond.

Prairie. A native plant community dominated by perennial grasses. *See also* steppe and short-grass prairie.

Precocial. The condition of being hatched or born in a relatively advanced state. *See also* altricial.

Promiscuity. A mating system in which at least one sex has multiple sexual partners and an absence of prolonged and individualized pair-bonding. *See also* polyandry, polygyny.

Rain shadow. The relatively dry area on the leeward side of a mountain, which is drier because of moisture primarily falling on the mountain's windward side.

Refugium (pl., refugia). An area where one or more species can survive during unfavorable environmental conditions, possibly for centuries or even millennia.

Relict. A population geographically isolated from a species' main range, suggesting that the latter's range was once broader or different and has since become fragmented.

Riparian. Associated with shorelines. Narrow gallery forests are a common type of riparian community in grassland habitats. *See also* gallery forest, riverine.

Riverine. Associated with rivers. *See also* gallery forest, riparian.

Rut (or rutting). Seasonal sexual activity, typical of many ungulates.

Sage scrub. A shrubsteppe community type dominated by sagebrush, usually with arid-adapted grasses, forbs, and other shrub types also present. *See also* shrubsteppe.

Saltatory. Refers to jumping (saltation).

Sand. Fine-grained natural materials larger than silt but smaller than gravel. Sandstone is a sedimentary rock composed mostly of hardened sand. *See also* silt.

Sandsage prairie. Mixed grass and shrubs (mostly sand sagebrush) developed over sandy substrates and with an abundance of sand-adapted species.

Savanna. A term now commonly used to describe any grassland community within which scattered trees occur. In typical savannas the distance between the trees is much greater than the average width of the individual tree canopies.

Saxicolous. Refers to dwelling among rocks.

Scansorial. Refers to climbing abilities.

Sedge. Herbaceous, grasslike plants having rounded or angular and solid stems, narrow, parallel-veined leaves that are often arranged in three ranks, and small flowers borne on spikes. *See also* grass.

Sensitive. A term used by the USDA Forest Service to identify species having downward trends in population numbers, population density, or "habitat capability."

Seral. Chronologically transitional or "successional" species or stages in community development. A successional sequence is called a sere. *See also* climax community, succession.

Sexual selection. A type of natural selection in which the evolution and maintenance of traits of one sex result from the social interactions that produce differential individual reproductive success. *See also* natural selection.

Short-grass prairie. Semiarid grasslands (sometimes called "steppes") dominated by short-stature perennial grasses. When shrubs are codominant with the grasses, "shrubsteppe" is a useful descriptive term. *See also* mixed-grass prairie, shrubsteppe, steppe.

Shrub. A woody plant that is typically less than twelve feet tall at maturity and usually has many above-ground stems. Shrublands are communities dominated by shrubs and having varied degrees of canopy cover. *See also* forest, tree, woodland.

Shrubsteppe. A semiarid plant community that is variably codominated by shrubs (typically various species of sagebrush in western North America) and low-stature grasses (often bunchgrasses). *See also* sagesteppe, steppe.

Siblings. Offspring of the same parents. Sibling species are two or more closely related species that resemble one another to a great degree.

Signals. Behaviors (visual, acoustic, tactile, and so on) that have become evolutionarily modified ("ritualized") to transmit information among members of a species' social group or to other organisms. Such behaviors are often also called "displays." *See also* advertising behavior, display.

Silt. Sedimentary materials of a size intermediate between sand and clay; becoming "loess" when wind-carried. *See also* loam, loess.

Site-fidelity (or site-tenacity). The social attachment of an individual to a specific site (nest site, territory, and so on) in successive seasons or years. *See also* philopatry.

Skylarking. A prolonged flight-song display, typical of many prairie-breeding birds.

Sod-forming grasses. Perennial grasses that grow and establish root systems tending to bind the soil substrate in a continuous sod rather than growing in a discontinuous, clumplike manner. *See also* bunchgrasses.

Songs. Vocalizations that tend to be acoustically complex, prolonged, and are often sex-specific as well as individually unique. They frequently are uttered only at particular seasons and may have both intra- and intersexual signal functions. *See also* calls, signals, vocalizations.

Species (abbreviated *sp.*, plural *spp.*). A "kind" of organism or, more technically, a population whose members are reproductively isolated from all other populations, but are capable of breeding freely among themselves. The term is unchanged in the plural. *See also* genus, Latin name.

Steppe. A Russian-based term for native short-grass communities. *See also* plains, short-grass prairie, shrubsteppe.

Strigid. A member of the typical owl family Strigidae.

Subspecies (abbreviated *ssp.*, plural *sspp.*). A geographically defined and recognizable (by morphology or sometimes by behavior) subdivision of a species.

Succession. The series of gradual plant and animal changes that occur in biotic communities over time as relatively temporary (successional or seral) taxa are sequentially replaced by others that are able to persist and reproduce for a more prolonged or even indefinite period. *See also* climax, community, seral.

Swamp. As used here, a wetland with standing or variably submerged woody vegetation. *See also* marsh, mire, lake.

Syrinx (adj., syringeal; plural, syringes). The vocal organ of birds, located at the posterior end of the trachea. The comparable mammalian structure is the larynx.

Tallgrass prairie. Perennial grasslands that are dominated by tall-stature grasses, often at least 6 feet high at maturity. *See also* mixed-grass prairie, short-grass prairie.

Territoriality. The advertisement and agonistic behaviors associated with territorial establishment and defense. *See also* advertising behavior, display, signals.

Territory. An area having resources that are defended or controlled by an animal against other of its species (intraspecific territories) or less often against those of other species (interspecific territories). *See also* advertising behavior, home range.

Thatch. Standing dead grass or similar herbaceous material. *See also* litter.

Thermal. A rising current of air caused by differential heating and expansion rates near the earth's surface.

Threatened. Descriptive of taxa that have declined and exist in small numbers, but are not yet so rare as to be classified as endangered. *See also* candidate, endangered.

Till. Unsorted glacial drift deposits. *See also* drift.

Trachea. The windpipe of land vertebrates, connecting the lungs with the oral cavity. *See also* syrinx.

Trait. A measurable phenotypic attribute (behavioral, structural, physiological), especially one that is at least in part genetically controlled.

Tree. A woody plant that is usually well above twelve feet tall at maturity and typically has a single main stem. *See also* shrub.

Trophic levels. The successive production and consumption levels in a food chain, including primary producers, primary consumers, secondary consumers, and so on. *See also* food chain.

Ungulate. A hoofed mammal; unguligrade locomotion involves using only the hooves for support.

Ventral. The lower or undersurface.

Vernacular name. The "common" or English-language name of an object or organism (usually a species) as opposed to its scientific or Latin name. *See also* Latin name.

Vernal pond. An ephemeral spring wetland produced by melting snow or early rains. *See also* lagoon, playa lake.

Vertebrates. Animals with backbones, including mammals, birds, reptiles, amphibians, and fish. *See also* invertebrates.

Vocalizations. Utterances, including both songs and calls, generated by (in birds) the syrinx or (in mammals and amphibians) the larynx and sometimes modified or modulated by other structures, such as the tongue, esophagus, or oral cavity. *See also* advertising behavior, calls, mechanical signals, songs.

Vole. A Norwegian-based term for any of a group of short-legged field mice, especially those of the meadow mouse genus *Microtus*.

Vulnerable. A term often used by state or other agencies to identify rare or declining species of possible future threatened or endangered conservation status. "Species of special concern" and "species in need of conservation" are similar terms. *See also* candidate, endangered, threatened.

Warm-season grasses. Grasses adapted to grow during warmer periods having greater water stress. Also called C_4 grasses in reference to an intermediate four-carbon molecule that is present during photosynthesis. *See also* cool-season grasses.

Water table. The upper surface of an underground zone of water-saturated soil or "aquifer."

Woodland. As used here, a partly wooded community in which the height of the trees is usually less than the distances between them so that the overhead canopy is discontinuous. The term is also often used in a more general sense, meaning any community largely dominated by trees. *See also* forest, savanna, tree.

Xeric. Desertlike or drought-adapted; e.g., "xerophilic" vegetation of deserts. *See also* mesic.

REFERENCES

The following references were chosen mainly for their relative accessibility, understandability, and relevance to the ecology and wildlife of the Great Plains states. No unpublished theses or dissertations were included, nor were in-house research reports of state conservation agencies.

GENERAL ECOLOGICAL, REGIONAL, AND COMPARATIVE AVIAN REFERENCES

Allen, D. 1967. *The Life of Prairies and Plains*. New York: McGraw Hill.
American Ornithologists' Union. 1998. *Checklist of North American Birds*. 7th ed. Washington, DC: American Ornithologists' Union.
Arnold, K., and K. Benson. In press. *Atlas of Texas Breeding Birds*. College Station: Texas A&M University Press. Preliminary distributional data are available online at *http://tbba.cbi.tamucc.edu/*.
Bailey, R. G. 1995. *Description of the Ecoregions of the United States*. 2d ed. Misc. Pub. No. 1391 (rev.). Washington, DC: USDA Forest Service.
Bailey, V. 1928. A biological survey of North Dakota, Bureau of Biological Survey, Washington, DC. *North American Fauna* 49:1–226.
Barbour, M. C., and W. D. Billings, eds. 1988. *North American Terrestrial Vegetation*. Cambridge: Cambridge University Press.
Benedict, R. A., P. W. Freeman, and H. Genoways. 1996. Prairie legacies—Mammals. In *Prairie Conservation: Preserving America's Most Endangered Ecosystem*, ed. F. B. Samson and F. L. Knopf. Pp. 149–66. Covelo, CA: Island Press.
Bent, A. C., ed. 1922–1968. Life histories of North American birds. *U.S. National Museum Bulletins* 107, 113, 121, 126, 130, 135, 142, 146, 162, 167, 170, 174, 176, 179, 191, 195, 196, 197, 203, 211, and 237.
Bleed, A., and C. Flowerday. 1989. *An Atlas of the Sand Hills*. 3d ed. Conservation and Survey Division Resource Atlas No. 5. Lincoln: University of Nebraska.

Brown, C. R., and M. B. Brown. 2001. Birds of the Cedar Point Biological Station. *Occasional Papers of the Cedar Point Biological Station,* Nos. 1.1–36. Lincoln: University of Nebraska.

Busby, W. H., and J. L. Zimmerman. 2001. *Kansas Breeding Bird Atlas.* Lawrence: University Press of Kansas.

Cable, T. T., S. Seltman, and K. J. Cook. 1997. *Birds of Cimarron National Grassland.* Gen. Tech. Rep. RM-GTR-281. Fort Collins, CO: USDA Forest Service, Rocky Mountain Forest and Range Experiment Station.

Chapman, K. A., M. K. Ziegenhagen, and A. Fischer. 1998. *Valley of Grass: Tallgrass Prairie and Parkland in the Red River Valley.* St. Cloud, MN: North Star Press.

Collins, J. T., ed. 1985. *Natural Kansas.* Lawrence: University Press of Kansas.

Collins, J. T., S. L. Collins, J. Horak, D. Mulhern, W. Busby, C. C. Freeman, and G. Wallace. 1995. *An Illustrated Guide to Endangered or Threatened Species in Kansas.* Lawrence: University Press of Kansas.

Cox, M. K., and W. L. Franklin. 1989. Terrestrial vertebrates of Scotts Bluff National Monument, Nebraska. *Southwestern Nat.* 49:597–613.

Crump, D. J., ed. 1984. *A Guide to Our Federal Lands.* Washington, DC: National Geographic Society.

Cushman, R. C., and S. R. Jones. 1988. *The Shortgrass Prairie.* Boulder, CO: Pruett.

Daubenmire, R. F. 1978. *Plant Geography: With Special Reference to North America.* New York: Academic Press.

Dechant, D. J., M. L. Sondreal, D. H. Johnson, L. D. Igl, C. M. Goldade, M. P. Nenneman, and B. R. Euliss. Effects of management practices on grassland birds, Northern Prairie Wildlife Research Center, Jamestown, ND. The following species summaries are available online at the Northern Prairie Research Station website, *<http://www.npwrc.usgov/resource/literatr/grasbird/grasbird.htm>:* American bittern, Baird's sparrow, burrowing owl, chestnut-collared longspur, clay-colored sparrow, dickcissel, field sparrow, ferruginous hawk, grasshopper sparrow, Henslow's sparrow, lark bunting, LeConte's sparrow, loggerhead shrike, marbled godwit, McCown's longspur, mountain plover, Nelson's sharp-tailed sparrow, Savannah sparrow, sedge wren, short-eared owl, Sprague's pipit, willet, and Wilson's phalarope.

Dobkin, D. S. 1994. *Conservation and Management of Neotropical Migrant Landbirds in the Northern Rockies and Great Plains.* Moscow: University of Idaho Press.

Dort, W., Jr., and J. K. Jones, Jr., eds. 1970. *Pleistocene and Recent Environments of the Central Great Plains.* Lawrence: University Press of Kansas.

Ducey, J. E. 1988. *Nebraska Birds: Breeding Status and Distribution.* Omaha, NE: Simmons-Boardman Books.

———. 1989. Birds of the Niobrara River Valley. *Trans. Nebraska Acad. Sci.* 17:37–60.

Faanes, C. E., and G. R. Lingle. 1995. Breeding birds of the Platte Valley of

Nebraska, Northern Prairie Wildlife Research Center, Jamestown, ND. Available online at <*http://www.npwrc.usgs.gov/resource/distr/birds/platte/ platte* (version 16JUL97).

Finch, D. M. 1992. *Threatened, Endangered, and Vulnerable Species of Terrestrial Vertebrates in the Rocky Mountain Region.* Gen. Tech. Rept. RM-215. Fort Collins, CO: USDA Forest Service, Rocky Mountain Forest and Range Experiment Station.

Froiland, S. G. 1978. *Natural History of the Black Hills.* Sioux Falls, SD: Center for Western Studies.

Gillian, S. W., and M. F. Carter. 2001. The shortgrass prairie. *Birding* 33(6):346–55.

Gress, R., and G. Potts. 1993. *Watching Kansas Wildlife.* Lawrence: University Press of Kansas.

Igl, L. D., and D. H. Johnson. 1997. Changes in breeding bird populations in North Dakota, 1967–1992–3. *Auk* 114:74–92.

Johnsgard, P. A. 1979. *Birds of the Great Plains: Breeding Species and Their Distribution.* Lincoln: University of Nebraska Press.

———. 1986. *Birds of the Rocky Mountains.* Boulder: Colorado Associated University Press.

———. 1995. *This Fragile Land: A Natural History of the Nebraska Sandhills.* Lincoln: University of Nebraska Press.

———. 2001a. *The Nature of Nebraska: Ecology and Biodiversity.* Lincoln: University of Nebraska Press.

———. 2001b. *Prairie Birds: Fragile Splendor in the Great Plains.* Lawrence: University Press of Kansas.

Jones, J. O. 1990. *Where the Birds Are: A Guide to All 50 States and Canada.* New York: William Morrow.

Kaul, R. B., G. E. Kantak, and S. P. Churchill. 1988. The Niobrara River Valley: A postglacial migration corridor and refugium of forest plants and animals in the grasslands of central North America. *Bot. Rev.* 54:44–81.

Knopf, F. L., and F. B. Samson, eds. 1997. *Ecology and Conservation of Great Plains Vertebrates.* New York: Springer.

Krue, J. 1992. *North Dakota Wildlife Viewing Guide.* Billings, MT: Falcon Press.

Küchler, A. W. 1966. *Potential Natural Vegetation of the Coterminous United States.* Spec. Pub. 36. New York: American Geographic Society.

Loope, D. B., and J. B. Swinehart. 2000. Thinking like a dune field: Geologic history in the Nebraska Sand Hills. *Great Plains Res.* 10:5–35.

Lynch, W. 1999. *Wild Birds across the Prairies.* Calgary: Fifth House.

McClure, H. E. 1966. Some observations of vertebrate fauna of the Nebraska Sandhills, 1941 through 1943. *Neb. Bird Review* 34:2–15.

Madson, J. 1993. *Tallgrass Prairie.* Helena, MT: Falcon Press.

———. 1995. *Where the Sky Began: Land of the Tallgrass Prairie.* Ames: Iowa State University Press.

Mollhoff, W. J. 2001. *The Nebraska Breeding Bird Atlas*. Lincoln: Nebraska Game and Parks Commission.

Mutel, C. F. 1989. *Fragile Giants: A Natural History of the Loess Hills*. Iowa City: University of Iowa Press.

National Geographic Staff. 1999. *Guide to Birdwatching Sites: Western North America*. Washington, DC: National Geographic Society.

Oklahoma Department of Conservation. 1994. *Oklahoma Wildlife Viewing Guide*. Billings, MT: Falcon Press.

Ortega, C. 1998. *Cowbirds and Other Brood Parasites*. Tucson: University of Arizona Press.

Peterson, R. A. 1993. *A Birdwatcher's Guide to the Black Hills and Adjacent Plains*. Vermillion, SD: P. C. Press.

————. 1995. *The South Dakota Breeding Bird Atlas*. Aberdeen: South Dakota Ornithologists' Union. Range maps available online through the Northern Prairie Wildlife Research Center website at *http://www.npwrc.usgs.gov/resource/othrdata/chekbird/chekbird.htm*.

Peterson, R. S., and C. S. Boyd. 1998. Ecology and Management of Sand Shinnery Communities: A Literature Review. Gen. Tech. Rep. RNRS-GTR-16. Fort Collins, CO: USDA Forest Service, Rocky Mountain Research Station.

Pettingill, O. S., Jr. 1981. *A Guide to Bird-finding West of the Mississippi*. 2d ed. New York: Oxford University Press.

Pettingill, O. S., Jr., and N. R. Whitney. 1965. *Birds of the Black Hills*. Pub. No. 1. Ithaca, NY: Cornell Laboratory of Ornithology.

Poole, A., and F. Gill, eds. 1992–2001. *The Birds of North America*. Philadelphia: Birds of North America. See especially the following accounts: #26–Western and Clark's Grebes, #210–Northern harrier, #298–Northern goshawk, #265–Swainson's hawk , #52–Red-tailed hawk, #172–Ferruginous hawk, #346–Prairie falcon, #515–Ruffed grouse, #425–Sage grouse, #354–Sharptailed grouse, #36–Greater prairie-chicken, #364–Lesser prairie-chicken, #31–Sandhill crane, #153–Whooping crane, #154–Snowy plover, #2–Piping plover, #211–Mountain plover, #580–Upland sandpiper, #492–Marbled godwit, #83–Wilson's phalarope, #116–Franklin's gull, #418–Yellow-billed cuckoo, #244–Greater roadrunner, #1–Barn owl, #165–Eastern screech-owl, #372–Great horned owl, #61–Burrowing owl, #62–Short-eared owl, #342–Northern saw-whet owl, #32–Common poorwill, #204–Ruby-throated hummingbird, #284–Lewis's woodpecker, #166–Northern flicker, #342–Scissortailed flycatcher, #452–Barn swallow, #231–Loggerhead shrike, #195–Horned lark, #486–Rock wren, #381–Eastern bluebird, #439–Sprague's pipit, #120–Clay-colored sparrow, #390–Brewer's sparrow, #488–Lark sparrow, #542–Lark bunting, #45–Savannah sparrow, #239–Grasshopper sparrow, #96–McCown's longspur, #288–Chestnut-collared longspur, #440–Northern cardinal, #4–Indigo bunting, #176–Bobolink, #271–Common grackle, #184–Red-winged blackbird, #192–Yellow-headed blackbird, #160–Eastern meadow-lark, #104–Western meadowlark, #47–Brown-headed cowbird, #256–Red crossbill.

Price, J., S. Droege, and A. Price. 1995. *The Summer Atlas of North American Birds.* New York: Academic Press.

Reichman, O. J. 1987. *Konza Prairie: A Tallgrass Natural History.* Lawrence: University Press of Kansas.

Rising, J. D. 1974. The status and faunal affinities of the summer birds of western Kansas. *University of Kansas Science Bull.* 50:347–88.

———. 1983. The Great Plains hybrid zones. *Current Ornithol.* 1:137–57.

———. 1996. *A Guide to the Identification and Natural History of the Sparrows of the United States and Canada.* San Diego, CA: Academic Press.

Samson, F., and F. Knopf. 1994. Prairie conservation in North America. *Bio-Science* 44:418–21.

———, eds. 1996. *Prairie Conservation: Conserving North America's Most Endangered Ecosystem.* Covelo, CA: Island Press.

Sauer, J. R., J. E. Hines, G. Gough, I. Thomas, and B. G. Peterson. 2000. The North American breeding bird survey results and analysis, 1966–2000, Patuxent Wildlife Research Center, Laurel, MD. Version 2000.1. Website: *http//www.mbr-pwrc.usgs.gov/bbs/.*

Seyffert, K. D. 2001. *Birds of the Texas Panhandle: Their Status, Distribution, and History.* College Station: Texas A&M University Press.

Sharpe, R. S., W. R. Silcock, and J. G. Jorgensen. 2001. *Birds of Nebraska: Their Distribution and Temporal Occurrence.* Lincoln: University of Nebraska Press.

Smith, J. N. M., T. L. Cook, S. I. Rothstein, S. K. Robinson, and S. G. Sealy, eds. 2000. *Ecology and Management of Cowbirds and Their Hosts.* Austin: University of Texas Press.

South Dakota Ornithologists' Union. 1991. *The Birds of South Dakota.* 2d ed. Aberdeen: South Dakota Ornithologists' Union.

Stewart, R. E. 1975. *Breeding Birds of North Dakota.* Fargo, ND: Tri-College Center for Environmental Studies.

Sutton, G. M. 1967. *Oklahoma Birds: Their Ecology and Distribution, with Comments on the Avifauna of the Southern Great Plains.* Norman: University of Oklahoma Press.

Thompson, M. C., and C. Ely. 1989, 1992. *Birds in Kansas.* 2 vols. Lawrence: University Press of Kansas.

Tomelleri, J. R., and M. E. Eberle. 1990. *Fishes of the Central United States.* Lawrence: University Press of Kansas.

Tremble, D. E. 1990. *The Geologic Story of the Great Plains.* Medora, ND: Theodore Roosevelt National Park and History Association.

Umber, H. 1988. The natural areas of North Dakota. *North Dakota Outdoors* 50(8):2–25.

U.S. Department of Interior, Geological Survey. 1970. *The National Atlas of the United States of America.* Washington, DC: Government Printing Office.

Weaver, J. E. 1954. *North American Prairie.* Lincoln, NE: Johnson Publishing.

Weaver, J. E., and F. W. Albertson. 1956. *The Grasslands of the Great Plains.* Lincoln, NE: Johnson.

Willig, M. R. 2001. Special feature: Prairie dogs. *J. Mammal.* 82:889–959.

Wishart, D. J., ed. In press. *Encyclopedia of the Great Plains.* Lincoln: University of Nebraska Press.

Wood, D. S., and G. D. Schnell. 1984. *Distributions of Oklahoma Birds.* Norman: University of Oklahoma Press.

Zimmerman, J. L. 1990. *Cheyenne Bottoms: Wetland in Jeopardy.* Lawrence: University Press of Kansas.

———. 1993. *The Birds of Konza: The Avian Ecology of the Tallgrass Prairie.* Lawrence: University Press of Kansas.

Zimmerman, J. L., and S. T. Patti. 1988. *A Guide to Bird-finding in Kansas and Western Missouri.* Lawrence: University Press of Kansas.

BIRDS (Species References)

Anderson, S. H., and J. R. Squires. 1997. *The Prairie Falcon.* Austin: University of Texas Press.

Austin, J. E., and A. D. Richert. 2001. A comprehensive review of observational and site evaluation data for migrant whooping cranes in the United States, 1943–99, U.S. Geological Survey, Northern Prairie Wildlife Research Center, Jamestown, ND.

Blair, C. L., and F. Schitoskey, Jr. 1982. Breeding biology and diet of the ferruginous hawk in South Dakota. *Wilson Bull.* 94:46–54.

Clark, R. J. 1975. A field study of the short-eared owl *Asio flammeus* Pontoppidan in North America. *Wildl. Monogr.* 47:1–67.

Davis, S. K., and S. G. Sealy. 1998. Nesting biology of the Baird's sparrow in southwestern Manitoba. *Wilson Bull.* 110:262–70.

Dunkle, F. W. 1977. Swainson's hawks on the Laramie Plains, Wyoming. *Auk* 94:65–71.

Granfors, D. A., K. E. Church, and L. M. Smith. 1996. Eastern meadowlarks nesting in rangelands and Conservation Reserve Program fields in Kansas. *J. Field Ornithol.* 67:222–35.

Graul, W. D. 1975. Breeding biology of the mountain plover. *Wilson Bull.* 87:6–31.

Hamerstrom, F. 1986. *Harrier: Hawk of the Marshes.* Washington, DC: Smithsonian Institution Press.

Harmeson, J. P. 1974. Breeding ecology of the dickcissel. *Auk* 91:348–59.

Higgins, K. F., and L. M. Kirsch. 1975. Some aspects of the breeding biology of the upland sandpiper in North Dakota. *Wilson Bull.* 87:96–101.

Higgins, K. F., L. M. Kirsch, M. R. Ryan, and R. B. Renken. 1979. Some ecological aspects of marbled godwits and willets in North Dakota. *Prairie Nat.* 11:115–18.

Jones, S. L., M. T. Green, and G. R. Geupel. 1998. Rare, little-known, and declining North American breeders. A closer look: Baird's sparrow. *Birding* 30:108–16.

Kagarise, C. 1979. Breeding biology of Wilson's phalarope in North Dakota. *Bird-Banding* 50:12–22.

Kirsch, L. M., and K. F. Higgins. 1976. Upland sandpiper nesting and management in North Dakota. *Wildl. Society Bull.* 4:16–20.

Lokemoen, J. T., and H. F. Duebbert. 1976. Ferruginous hawk nesting ecology and raptor populations in northern South Dakota. *Condor* 78:464–70.

Porter, D. K., M. A. Strong, J. B. Giezentanner, and R. A. Ryder. 1975. Nest ecology, productivity, and growth of the loggerhead shrike on the shortgrass prairie. *Southwest. Nat.* 19:429–36.

Sibley, C. G., and L. L. Short, Jr. 1964. Hybridization in the orioles of the Great Plains. *Condor* 66:130–50.

Sibley, C. G., and D. A. West. 1959. Hybridization in the rufous-sided towhees of the Great Plains. *Auk* 76:326–38.

Winter, M. 1999a. Nesting biology of dickcissels and Henslow's sparrows in southwestern Missouri prairie fragments. *Wilson Bull.* 111:515–25.

———. 1999b. Relationship of fire history to territory size, breeding density, and habitat of Baird's sparrow in North Dakota. *Studies in Avian Biology* 19:171–77.

Zimmerman, J. L. 1988. Breeding season habitat selection by the Henslow's sparrow (*Ammodramus henslowii*) in Kansas. *Wilson Bull.* 100:17–24.

MAMMALS (Species and Regional References)

Beckoff, M., ed. 1978. *Coyotes: Biology, Behavior, and Management.* New York: Academic Press.

Bee, J. W., G. E. Glass, R. S. Hoffmann, and R. R. Patterson. 1981. Mammals in Kansas. *Pub. Ed. Ser., Mus. Nat. Hist., Univ. Kansas* 7:1–300.

Benedict, R. A., P. W. Freeman, and H. H. Genoways. 1996. Prairie legacies—Mammals. In *Prairie Conservation: Conserving North America's Most Endangered Ecosystem,* ed. F. B. Samson and F. L. Fritz. Pp. 149–66. Covelo, CA: Island Press.

Benedict, R. A., H. H. Genoways, and P. W. Freeman. 2000. Shifting distributional patterns of mammals in Nebraska. *Trans. Nebraska Acad. Sci.* 26:55–84.

Bowles, J. B. 1975. Distribution and biogeography of the mammals of Iowa. *Spec. Publ. Mus., Texas Tech University* 9:1–184.

Burt, W. H., and R. P. Grossenheider. 1976. *A Field Guide to the Mammals.* 3d ed. Boston: Houghton Mifflin.

Byers, J. A. 1997. *American Pronghorn: Social Adaptation and the Ghosts of Predators Past.* Chicago: University of Chicago Press.

Caire, W., J. D. Taylor, B. P. Glass, and M. A. Mares. 1989. *Mammals of Oklahoma.* Norman: University of Oklahoma Press.

Campbell, T. M., III, and T. W. Clark. 1981. Colony characteristics and vertebrate associates of white-tailed and black-tailed prairie dogs in Wyoming. *Amer. Midl. Nat.* 105:269–75.

Chapman, J. A., and C. A. Feldhammer, eds. 1982. *Wild Mammals of North America.* Baltimore: Johns Hopkins University Press.

Choate, J. R., R. K. Jones, Jr., and C. Jones. 1994. *Handbook of Mammals of the South-central States.* Baton Rouge: Louisiana State University Press.

Choate, L. L. 1997. The mammals of the Llano Estacado. *Spe. Publ. Mus., Texas Tech Univ.* 40:1–240.

Clark, T. W., ed. 1986. The black-footed ferret. *Great Basin Naturalist Memoirs* 8:1–308.

Clark, T. W., T. M. Campbell III, D. C. Socha, and D. E. Casey. 1982. Prairie dog colony attributes and associated vertebrate species. *Great Basin Nat.* 42:572–82.

Clark, T. W., and M. R. Stromberg, 1987. *Mammals in Wyoming.* Publication Education Series no. 10. Lawrence: University of Kansas Museum of Natural History.

Cockrum, E. L. 1952. Mammals of Kansas. *Publ. Mus. Nat. Hist., Univ. Kansas* 7:1–303.

Costello, D. F. 1970. *The World of the Prairie Dog.* New York: J. P. Lippincott.

Crabb, W. D. 1948. The ecology and management of the spotted skunk in Iowa. *Ecol. Monogr.* 18:201–32.

Davis, W. B., and D. J. Schmidly. 1995. *The Mammals of Texas.* Austin: University of Texas Press.

Demaris, S., and P. R. Krausman, eds. 2000. *Ecology and Management of Large Mammals in North America.* Upper Saddle River, NJ: Prentice Hall.

Desmond, M. J., J. A. Savidge, and K. M. Eskridge. 2000. Correlations between burrowing owl and black-tailed prairie dog declines. *J. Wildl. Mgmt.* 64:1067–75.

Downhower, J. F., and E. R. Hall. 1966. The pocket gopher in Kansas. *Misc. Publ. Mus. Nat. Hist., Univ. Kansas* 44:1–32.

Eisenberg, J. E. 1963. The behavior of heteromyid rodents. *Univ. California Publ. Zool.* 69:1–114.

Findley, J. S. 1987. *The Natural History of New Mexico Mammals.* Albuquerque: University of New Mexico Press.

Findley, J. S., A. H. Harris, D. E. Wilson, and C. Jones. 1975. *Mammals of New Mexico*. Albuquerque: University of New Mexico Press.

Fitzgerald, J. P., C. A. Meanley, and D. M. Armstrong. 1994. *Mammals of Colorado*. Niwot: University of Colorado Press.

Forbes, R. B. 1964. Some aspects of the life history of the silky pocket mouse, *Perognathus flavus. Amer. Midl. Nat.* 72:290–308.

Forsman, K. R. 2001. *The Wild Mammals of Montana*. Lawrence, KS: American Society of Mammalogists. Website: *asm@allenpress.com*.

Forsyth, A. 1999. *Mammals of North America: Temperate and Arctic Regions*. Buffalo, NY: Firefly Books.

Garner, H. W. 1974. Population dynamics, reproduction, and activities of the kangaroo rat, *Dipodomys ordii*, in western Texas. *Graduate Studies, Texas Tech. Univ.* 7:1–28.

Geist, V. 1996. *Buffalo Nation: History and Legends of the North American Bison*. Stillwater, MN: Voyageur Press.

Goertz, J. W. 1963. Some biological notes on the plains harvest mouse. *Proc. Oklahoma Acad. Sci.* 43:123–25.

Haberman, C. G., and E. D. Fleharty. 1972. Natural history notes on Franklin's ground squirrel in Boone County, Nebraska. *Trans. Kansas Acad. Sci.* 74:76–80.

Hall, E. R. 1965. Handbook of mammals of Kansas. *Misc. Publ. Mus. Nat. Hist., Univ. Kansas* 7:1–303.

———. 1981. *The Mammals of North America*. 2 vols. New York: John Wiley and Sons.

Hazard, E. B. 1982. *The Mammals of Minnesota*. Minneapolis: University of Minnesota Press.

Higgins, K. F., E. D. Stukel, J. M. Goulet, and D. C. Backlund. 2000. *Wild Mammals of South Dakota*. Pierre: South Dakota Department of Game, Fish, and Parks.

Hoffman, R. S., and D. L. Pattie. 1968. *A Guide to Montana Mammals*. Missoula: University of Montana.

Jameson, E. W., Jr. 1947. Natural history of the prairie vole (mammalian genus *Microtus*). *Misc. Publ. Mus. Nat. Hist., Univ. Kansas* 1:125–51.

Jones, J. K., Jr. 1964. Distribution and taxonomy of mammals of Nebraska. *Publ. Mus. Nat. Hist., Univ. Kansas* 16:1–356.

Jones, J. K., Jr., D. M. Armstrong, and J. R. Choate. 1985. *Guide to Mammals of the Plains States*. Lincoln: University of Nebraska Press.

Jones, J. K., Jr., D. M. Armstrong, R. S. Hoffmann, and C. Jones. 1983. *Mammals of the Northern Great Plains*. Lincoln: University of Nebraska Press.

Jones, J. K., Jr., and E. C. Birney. 1988. *Handbook of Mammals of the North-central States*. Minneapolis: University of Minnesota Press.

Kilgore, D. L., Jr. 1969. An ecological study of the swift fox (*Vulpes velox*) in the Oklahoma panhandle. *Amer. Midl. Nat.* 81:512–34.

King, J. A. 1955. Social behavior, social organization, and population dynamics of a black-tailed prairie dog town in the Black Hills of South Dakota. *Contrib. Lab. Vert. Biol., Univ. of Mich.* 67:1–123.

———. 1968. Biology of *Peromyscus*. *American Society of Mammalogists Special Publication No. 2*:1–593.

Kitchen, D. W. 1974. Social behavior and ecology of the pronghorn. *Wildl. Monogr.* 38:1–96.

Long, C. A. 1965. The mammals of Wyoming. *Publ. Mus. Nat. Hist., Univ. Kansas* 14:493–758.

McCarley, H. 1966. Annual cycle, population dynamics, and adaptive behavior of *Citellus tridecemlineatus*. *J. Mammal.* 47:294–316.

McMillan, B. R., D. W. Kaufman, G. A. Kaufman, and R. S. Mattack. 1997. Mammals of Konza Prairie: New observations and an updated species list. *Prairie Nat.* 29:263–71.

Miller, B., G. Ceballos, and R. Reading. 1994. The prairie dog and biotic diversity. *Conserv. Biol.* 8:677–81.

Murie, J., and C. D. Michener, eds. 1984. *The Biology of Ground-dwelling Squirrels*. Lincoln: University of Nebraska Press.

Novak, M., J. A. Baker, M. E. Obbard, and B. Malloch, eds. 1987. *Wild Furbearer Management and Conservation in North America*. Toronto: Ontario Ministry of Natural Resources.

Orr, R. T. 1970. *Mammals of North America*. New York: Doubleday.

Reed, K. M., and J. R. Choate. 1986. Natural history of the plains pocket mouse in agriculturally disturbed sandsage prairie. *Prairie Nat.* 18:79–90.

Roe, F. G. 1951. *The North American Buffalo: A Critical Study of a Species in Its Wild State*. Toronto: University of Toronto Press.

Schwartz, C. W., and E. R. Schwartz. 1981. *The Wild Mammals of Missouri*. 2d ed. Columbia: University of Missouri Press.

Seabloom, R. W., and P. W. Theisen. 1990. Breeding biology of the black-tailed prairie dog in North Dakota. *Prairie Nat.* 22:65–74.

Seal, U., E. T. Thorne, M. Bogan, and S. Anderson, eds. 1989. *Conservation Biology and the Black-footed Ferret*. New Haven, CT: Yale University Press.

Sealander, J. A. 1979. *A Guide to Arkansas Mammals*. Conway, AR: River Road Press.

Smith, R. E. 1967. Natural history of the prairie dog in Kansas. *Misc. Publ. Mus. Nat. Hist., Univ. Kansas* 16:1–36.

Steele, M. A., and J. L. Koprowski. 2001. *North American Tree Squirrels*. Washington, DC: Smithsonian Institution Press.

Stromberg, M. R., and M. S. Boyce. 1986. Systematics and conservation of the swift fox, *Vulpes velox*, in North America. *Biol. Conserv.* 35:97–110.

Tamarin, R. H., ed. 1985. Biology of New World *Microtus*. *American Society of Mammalogists Special Publication No. 8*:1–893.

Turner, R. W. 1974. Mammals of the Black Hills of South Dakota and Wyoming. *Misc. Publ. Mus. Nat. Hist., Univ. Kansas* 60:1–178.

Tyler, J. D. 1970. Vertebrates in a prairie dog town. *Proc. Oklahoma Acad. Sci.* 50:110–13.

Verts, B. J. 1967. *The Biology of the Striped Skunk.* Urbana: University of Illinois Press.

Wemmer, C., ed. 1987. *Biology and Management of the Cervidae.* Washington, DC: Smithsonian Institution Press.

Whitaker, J. O., Jr. 1996. *National Audubon Society Field Guide to North American Mammals.* New York: A. A. Knopf.

Whitaker, J. O., Jr., and W. J. Hamilton, Jr. 1998. *Mammals of the Eastern United States.* 3d ed. Ithaca, NY: Cornell University Press.

Wilson, D. E., and S. Ruff. 1999. *The Smithsonian Book of North American Mammals.* Washington, DC: Smithsonian Institution Press.

Young, S. P. 1944. *The Wolves of North America.* Washington, DC: American Wildlife Institute.

REPTILES AND AMPHIBIANS (Species and Regional References)

Anderson, P. 1965. *The Reptiles of Missouri.* Columbia: University of Missouri Press.

Ballinger, R. E., J. E. Lynch, and P. H. Cole. 1979. Distribution and natural history of amphibians and reptiles in western Nebraska, with ecological notes on the herptiles of Arapaho Prairie. *Prairie Nat.* 22:65–74.

Ballinger, R. E., J. W. Meeker, and M. Thies. 2000. A checklist and distribution maps of the amphibians and reptiles of South Dakota. *Trans. Nebr. Acad. Sci.* 26:29–46.

Behler, J. L., and F. W. King. 1996. *National Audubon Society Field Guide to North American Reptiles and Amphibians.* New York: A. A. Knopf.

Benedict, R. 1996. Snappers, soft-shells, and stinkpots: The turtles of Nebraska. *Museum Notes* (University of Nebraska State Museum, Lincoln) 96:1–4.

Black, J. H., and G. Sievert. 1989. *A Field Guide to the Amphibians of Oklahoma.* Oklahoma City: Oklahoma Department of Wildlife Conservation.

Caldwell, J. P., and J. T. Collins. 1981. *Turtles in Kansas.* Lawrence, KS: AMS Publishing.

Collins, J. T. 1982. *Amphibians and Reptiles in Kansas.* Publication Education Series, no. 13. Lawrence: University of Kansas Museum of Natural History.

Collins, J. T., and S. L. Collins. 1991. *Reptiles and Amphibians of the Cimarron National Grasslands.* Elkhart, KS: U.S. Forest Service.

Conant, R. 1998. *A Field Guide to the Reptiles and Amphibians: Eastern and Central North America.* 3d ed. Boston: Houghton Mifflin.

Dixon, J. R. 2000. *Amphibians and Reptiles of Texas.* 2d ed. College Station: Texas A&M University Press.

Duellman, W. E., and L. Trueb. 1986. *Biology of Amphibians.* New York: McGraw-Hill.

Ernst, C. H., J. E. Lovich, and R. W. Barbour. 1994. *Turtles of the United States and Canada.* Washington, DC: Smithsonian Institution Press.

Fisher, T., D. Backlund, K. W. Higgins, and D. Naugle. 1999. *A Field Guide to South Dakota Amphibians.* Agricultural Extension Bulletin B733. Brookings: South Dakota State University.

Freeman, P. 1989. Amphibians and reptiles. In *An Atlas of the Sand Hills,* ed. A. Bleed and C. Flowerday. Pp. 157–60. Conservation and Survey Division Resource Atlas No. 5. Lincoln: University of Nebraska.

Garrett, J. M., and D. G. Barker. 1987. *A Field Guide to Reptiles and Amphibians of Texas.* Austin: Texas Monthly Press.

Hoberg, T. D., and C. T. Gause. 1992. Reptiles and amphibians of North Dakota. *North Dakota Outdoors* 55(1):7–20.

Holycross, A. 1995. Serpents of the Sandhills. *Nebraskaland* 73(6):28–35.

Hudson, G. E. 1958. *The Amphibians and Reptiles of Nebraska. Nebraska Conservation Bulletin 24.* Lincoln: University of Nebraska Conservation and Survey Division.

Jones, S. M., and R. E. Ballinger. 1987. Comparative life histories of *Holbrookia maculata* and *Sceloperus undulatus* in western Nebraska. *Ecology* 68:1828–36.

Klauber, L. M. 1972. *Rattlesnakes: Their Habits, Life Histories, and Influence on Mankind.* 2d ed. 2 vols. Berkeley: University of California Press.

Legler, J. M. 1960. Natural history of the ornate box turtle, *Terrapene ornata ornata. Publ. Mus. Nat. Hist., Univ. Kansas* 11:527–669.

Lynch, J. D. 1985. Annotated checklist of the amphibians and reptiles of Nebraska. *Trans. Nebraska Acad. Sci.* 13:33–57.

Oliver, J. A. 1955. *The Natural History of North American Amphibians and Reptiles.* Princeton, NJ: Van Nostrand.

Rossman, D. A., N. B. Ford, and R. A. Seigel. 1996. *The Garter Snakes: Evolution and Ecology.* Norman: University of Oklahoma Press.

Sievert, G., and L. Sievert. 1988. *A Field Guide to the Reptiles of Oklahoma.* Oklahoma City: Oklahoma Department of Wildlife Conservation.

Smith, H. M. 1956. Handbook of amphibians and reptiles of Kansas. *Publ. Mus. Nat. Hist., Univ. Kansas* 9:1–356.

Smith, H. M., and E. D. Brodie, Jr. 1982. *Reptiles of North America: A Guide to Field Identification.* New York: Golden Press.

Stebbens, R. C. 1978. *A Field Guide to Western Reptiles and Amphibians.* Boston: Houghton Mifflin.

Tennant, A. 1984. *The Snakes of Texas.* Austin: Texas Monthly Press.

Thompson, S., and D. Backlund. 1998. *South Dakota Snakes: A Guide to Snake Identification*. Pierre: South Dakota Department of Game, Fish, and Parks.

Tyning, T. F. 1990. *A Guide to Amphibians and Reptiles*. Boston: Little, Brown.

Webb, R. G. 1970. *Reptiles of Oklahoma*. Norman: University of Oklahoma Press.

Wheeler, G. C., and J. Wheeler. 1966. *The Amphibians and Reptiles of North Dakota*. Grand Forks: University of North Dakota Press.

Wright, A. H., and A. A. Wright. 1949. *Handbook of Frogs and Toads*. Ithaca, NY: Comstock Publishing.

———. 1957. *Handbook of Snakes of the United States and Canada*. 2 vols. Ithaca, NY: Comstock Publishing.

SPECIES INDEX

Pages with species illustrations are shown in italics.